Thermal Engineering Handbook

Thermal Engineering Handbook

Peter Banks

CLANRYE
INTERNATIONAL
www.clanryeinternational.com

Clanrye International,
750 Third Avenue, 9th Floor,
New York, NY 10017, USA

ISBN: 978-1-64726-139-9

Cataloging-in-Publication Data

Thermal engineering handbook / Peter Banks.
p. cm.
Includes bibliographical references and index.
ISBN: 978-1-64726-139-9
1. Heat engineering. 2. Power (Mechanics). 3. Thermodynamics. I. Banks, Peter.
TJ255 .T44 2022
621.402--dc23

For information on all Clanrye International publications
visit our website at www.clanryeinternational.com

CLANRYE INTERNATIONAL

Table of Contents

Preface

The purpose of this book is to help students understand the fundamental concepts of this discipline. It is designed to motivate students to learn and prosper. I am grateful for the support of my colleagues. I would also like to acknowledge the encouragement of my family.

Thermal engineering is a sub-discipline of mechanical engineering that focuses on the movement and transfer of heat energy. The energy is transformed between two mediums. It can also be transferred into other forms of energy. Thermal engineering makes use of thermodynamics, which is a branch of physics that deals with heat and temperature. It involves the process of converting the generated energy from thermal sources into mechanical, chemical and electrical energy. Thermofluids is an associated field of thermal engineering. It draws on concepts from thermodynamics as well as thermal engineering. This book presents the complex subject of thermal engineering in the most comprehensible and easy to understand language. It explores all the important aspects of thermal engineering in the present day scenario. This book will serve as a valuable source of reference for those interested in this field.

A foreword for all the chapters is provided below:

Chapter - What is Thermal Engineering?

The branch of engineering that deals with the movement of heat energy and transfer is known as thermal engineering. Laws of thermodynamics, enthalpy and entropy are some of the important concepts of thermal engineering. This is an introductory chapter which will introduce briefly all these major concepts of thermal engineering.

Chapter - Steam Power Cycles

Many different types of cycles are used to operate a steam power cycle. Some of these cycles are Rankine cycle, Carnot cycle, binary cycle, Diesel cycle, Atkinson cycle, Brayton cycle, etc. The topics elaborated in this chapter will help in gaining a better perspective about the various operations of these steam power cycles.

Chapter - Steam Generators

A power generation device which is used to generate steam by applying heat energy to water is referred to as steam generator. They can be categorized into water tube boilers and fire tube boilers. This chapter has been carefully written to provide an easy understanding of these classifications of steam generators as well as their working.

Chapter - Understanding Draught

Draught is defined as the pressure difference in the boiler system due to which the gas flows from one point to another. Artificial draught, induced draught, balanced draught and steam jet draught are some of the different types of draught. This chapter discusses in detail these types of draught.

Chapter - Steam Nozzle and Injector

A duct through which the velocity of a fluid increases due to drop in pressure is called a steam nozzle. A steam injector is a passive jet pump which produces higher pressure water than the inlet fluid pressure without using external power source. This chapter closely examines the key concepts of steam nozzle and injector to provide an extensive understanding of the subject.

Peter Banks

1
What is Thermal Engineering?

The branch of engineering that deals with the movement of heat energy and transfer is known as thermal engineering. Laws of thermodynamics, enthalpy and entropy are some of the important concepts of thermal engineering. This is an introductory chapter which will introduce briefly all these major concepts of thermal engineering.

Thermal Engineering is a specialised sub-discipline of Mechanical Engineering that deals exclusively with heat energy and its transfer between not only different mediums, but also into other usable forms of energy. A thermal engineer will be armed with the expertise to design systems and process to convert generated energy from various thermal sources into chemical, mechanical or electrical energy depending on the task at hand. Obviously, all thermal engineers are experts in all aspects of heat transfer.

Many process plants (basically somewhere where some raw material or resource is converted into something useful, e.g. power plants, oil refineries, plastic manufacturing plants, etc.) contain countless components and systems which have to be designed in terms of their heat transfer; it is particularly important to ensure that not too much heat is retained so the component or process is not disrupted. Conversely, some processes or systems are designed to use heat to their advantage and a thermal engineer must make sure enough heat is generated and used wisely (i.e. sustainably).

Thermal Engineers must also know about the economics of the components and processes they design to make sure they not only provide an improvement over the existing solutions, but also don't lose the company money. thermal engineers are not limited in areas of specialisation and can work in numerous fields. Below is only a brief example of areas a thermal engineer can work in:

- Heating, Ventilation and Air Conditioning (HVAC) systems in small and large-scale residential, commercial or industrial buildings.
- Renewable energy systems.
- Military and defence equipment.
- Electronics and electrical component and systems.

- Aerospace components.
- Boiler, heat exchanger, and pump design, amongst others.

Common industries that regularly employ thermal engineers includes power companies, the automotive industry and commercial construction. While thermal engineers will generally spend most of their time working in an office they are often required to travel to the site of their current project.

Heat

Heat is the transfer of kinetic energy from one medium or object to another, or from an energy source to a medium or object. Such energy transfer can occur in three ways: radiation, conduction, and convection.

The standard unit of heat in the International System of Units (SI) is the calorie (cal), which is the amount of energy transfer required to raise the temperature of one gram of pure liquid water by one degree Celsius, provided the water temperature is higher than the freezing point and lower than the boiling point. Sometimes the kilocalorie (kcal) is specified as a unit of heat; 1 kcal = 1000 cal. (This is the so-called diet calorie.) Less often, the British thermal unit (Btu) is used. This is the amount of heat required to raise the temperature of one pound of pure liquid water by one degree Fahrenheit.

An example of heat by radiation is the effect of infrared (IR) energy as it strikes a surface. IR is an electromagnetic field capable of transfering energy from a source, such as a fireplace, to a destination, such as the surfaces within a room. Radiation does not require an intervening medium; it can occur through a vacuum. It is responsible for the warming of the Earth by the sun.

Heat by conduction takes place when two material media or objects are in direct contact, and the temperature of one is higher than the temperature of the other. The temperatures tend to equalize; thus the heat conduction consists of a transfer of kinetic energy from the warmer medium to the cooler one. An example is the immersion of a chilled human body in a hot bath.

Heat by convection occurs when the motion of a liquid or gas carries energy from a warmer region to a cooler region. A good example of convection is the tendency of warm air to rise and cool air to fall, equalizing the air temperature inside a room containing a hot stove. Heat convection (along with conduction) is believed to take place inside the Earth, transfering kinetic energy from the inner core through the outer core and mantle to the crust. In this situation, the outer core and the mantle behave like liquids over long periods of time.

Thermal Conduction

Thermal conduction is the transfer of heat internal energy by microscopic collisions of particles and movement of electrons within a body. The microscopically colliding particles, that include molecules, atoms and electrons, transfer disorganized microscopic kinetically and potential energy, jointly known as internal energy. Conduction takes place in all phases of including solids, liquids, gases and waves. The rate at which energy is conducted as heat between two bodies is a function of the temperature difference temperature gradient between the two bodies and the properties of the conductive through which the heat is transferred.

Heat spontaneously flows from a hotter to a colder body. For example, heat is conducted from the hotplate of an electric stove to the bottom of a saucepan in contact with it. In the absence of an external driving energy source to the contrary, within a body or between bodies, temperature differences decay over time, and thermal equilibrium is approached, temperature becoming more uniform.

In conduction, the heat flow is within and through the body itself. In contrast, in heat transfer by thermal radiation, the transfer is often between bodies, which may be separated spatially. Also possible is transfer of heat by a combination of conduction and thermal radiation. In convection, internal energy is carried between bodies by a moving material carrier. In solids, conduction is mediated by the combination of vibrations and collisions of molecules, of propagation and collisions of phonons, and of diffusion and collisions of free electrons. In gases and liquids, conduction is due to the collisions and diffusion of molecules during their random motion. Photons in this context do not collide with one another, and so heat transport by electromagnetic radiation is conceptually distinct from heat conduction by microscopic diffusion and collisions of material particles and phonons. But the distinction is often not easily observed, unless the material is semi-transparent.

In the engineering sciences, heat transfer includes the processes of thermal radiation, convection, and sometimes mass transfer. Usually, more than one of these processes occurs in a given situation. The conventional symbol for thermal conductivity is k.

On a microscopic scale, conduction occurs within a body considered as being stationary; this means that the kinetic and potential energies of the bulk motion of the body are separately accounted for. Internal energy diffuses as rapidly moving or vibrating atoms and molecules interact with neighbouring particles, transferring some of their microscopic kinetic and potential energies, these quantities being defined relative to the bulk of the body considered as being stationary. Heat is transferred by conduction when adjacent atoms or molecules collide, or as several electrons move backwards and forwards from atom to atom in a disorganized way so as not to form a macroscopic electric current, or as photons collide and scatter. Conduction is the most significant means of heat transfer within a solid or between solid objects in thermal contact. Conduction is greater in solids because the network of relatively close fixed spatial relationships between atoms helps to transfer energy between them by vibration.

Thermal contact conductance is the study of heat conduction between solid bodies in contact. A temperature drop is often observed at the interface between the two surfaces in contact. This phenomenon is said to be a result of a thermal contact resistance existing between the contacting surfaces. Interfacial thermal resistance is a measure of an interface's resistance to thermal flow. This thermal resistance differs from contact resistance, as it exists even at atomically perfect interfaces. Understanding the thermal resistance at the interface between two materials is of primary significance in the study of its thermal properties. Interfaces often contribute significantly to the observed properties of the materials.

The inter-molecular transfer of energy could be primarily by elastic impact, as in fluids, or by free electron diffusion, as in metals, or phonon vibration, as in insulators. In insulators, the heat flux is carried almost entirely by phonon vibrations.

Metals (e.g., copper, platinum, gold, etc.) are usually good conductors of thermal energy. This is due to the way that metals bond chemically: metallic bonds (as opposed to covalent or ionic bonds) have free-moving electrons that transfer thermal energy rapidly through the metal. The *electron fluid* of a conductive metallic solid conducts most of the heat flux through the solid. Phonon flux is still present, but carries less of the energy. Electrons also conduct electric current through conductive solids, and the thermal and electrical conductivities of most metals have about the same ratio. A good electrical conductor, such as copper, also conducts heat well. Thermoelectricity is caused by the interaction of heat flux and electric current. Heat conduction within a solid is directly analogous to diffusion of particles within a fluid, in the situation where there are no fluid currents.

In gases, heat transfer occurs through collisions of gas molecules with one another. In the absence of convection, which relates to a moving fluid or gas phase, thermal conduction through a gas phase is highly dependent on the composition and pressure of this phase, and in particular the mean free path of gas molecules relative to the size of the gas gap, as given by the Knudsen number K_n.

To quantify the ease with which a particular medium conducts, engineers employ the thermal conductivity, also known as the conductivity constant or conduction coefficient, *k*. In thermal conductivity, *k* is defined as "the quantity of heat, *Q*, transmitted in time (*t*) through a thickness (*L*), in a direction normal to a surface of area (*A*), due to a temperature difference (ΔT)". Thermal conductivity is a material *property* that is primarily dependent on the medium's phase, temperature, density, and molecular bonding. Thermal effusivity is a quantity derived from conductivity, which is a measure of its ability to exchange thermal energy with its surroundings.

Steady-state Conduction

Steady state conduction is the form of conduction that happens when the temperature difference(s) driving the conduction are constant, so that (after an equilibration time), the spatial distribution of temperatures (temperature field) in the conducting object does not change any further. Thus, all partial derivatives of temperature *with respect to space* may either be zero or have nonzero values, but all derivatives of temperature at any point *with respect to time* are uniformly zero. In steady state conduction, the amount of heat entering any region of an object is equal to amount of heat coming out (if this were not so, the temperature would be rising or falling, as thermal energy was tapped or trapped in a region).

For example, a bar may be cold at one end and hot at the other, but after a state of steady state conduction is reached, the spatial gradient of temperatures along the bar does not change any further, as time proceeds. Instead, the temperature remains constant at any given cross-section of the rod normal to the direction of heat transfer, and this temperature varies linearly in space in the case where there is no heat generation in the rod.

In steady state conduction, all the laws of direct current electrical conduction can be applied to "heat currents". In such cases, it is possible to take "thermal resistances" as the analog to electrical resistances. In such cases, temperature plays the role of voltage, and heat transferred per unit time (heat power) is the analog of electric current. Steady state systems can be modelled by networks of such thermal resistances in series and in parallel, in exact analogy to electrical networks of resistors.

Transient Conduction

During any period in which temperatures changes *in time* at any place within an object, the mode of thermal energy flow is termed *transient conduction*. Another term is "non steady-state" conduction, referring to time-dependence of temperature fields in an object. Non-steady-state situations appear after an imposed change in temperature at a boundary of an object. They may also occur with temperature changes inside an object, as a result of a new source or sink of heat suddenly introduced within an object, causing temperatures near the source or sink to change in time.

When a new perturbation of temperature of this type happens, temperatures within the system change in time toward a new equilibrium with the new conditions, provided that these do not change. After equilibrium, heat flow into the system once again equals the heat flow out, and temperatures at each point inside the system no longer change. Once this happens, transient conduction is ended, although steady-state conduction may continue if heat flow continues.

If changes in external temperatures or internal heat generation changes are too rapid for the equilibrium of temperatures in space to take place, then the system never reaches a state of unchanging temperature distribution in time, and the system remains in a transient state.

An example of a new source of heat "turning on" within an object, causing transient conduction, is an engine starting in an automobile. In this case, the transient thermal conduction phase for the entire machine is over, and the steady state phase appears, as soon as the engine reaches steady-state operating temperature. In this state of steady-state equilibrium, temperatures vary greatly from the engine cylinders to other parts of the automobile, but at no point in space within the automobile does temperature increase or decrease. After establishing this state, the transient conduction phase of heat transfer is over.

New external conditions also cause this process: for example the copper bar in the example steady-state conduction experiences transient conduction as soon as one end is subjected to a different temperature from the other. Over time, the field of temperatures inside the bar reach a new steady-state, in which a constant temperature gradient along the bar is finally set up, and this gradient then stays constant in space. Typically, such a new steady state gradient is approached exponentially with time after a new temperature-or-heat source or sink, has been introduced. When a "transient conduction" phase is over, heat flow may still continue at high power, so long as temperatures do not change.

An example of transient conduction that does not end with steady-state conduction, but rather no conduction, occurs when a hot copper ball is dropped into oil at a low temperature. Here, the temperature field within the object begins to change as a function of time, as the heat is removed from the metal, and the interest lies in analyzing this spatial change of temperature within the object over time, until all gradients disappear entirely (the ball has reached the same temperature as the oil). Mathematically, this condition is also approached exponentially; in theory it takes infinite time, but in practice it is over, for all intents and purposes, in a much shorter period. At the end of this process with no heat sink but the internal parts of the ball (which are finite), there is no steady state heat conduction to reach. Such a state never occurs in this situation, but rather the end of the process is when there is no heat conduction at all.

The analysis of non steady-state conduction systems is more complex than that of steady-state systems. If the conducting body has a simple shape, then exact analytical mathematical expressions and solutions may be possible. However, most often, because of complicated shapes with varying thermal conductivities within the shape (i.e., most complex objects, mechanisms or machines in engineering) often the application of approximate theories is required, and/ or numerical analysis by computer. One popular graphical method involves the use of Heisler Charts.

Occasionally, transient conduction problems may be considerably simplified if regions of the object being heated or cooled can be identified, for which thermal conductivity is very much greater than that for heat paths leading into the region. In this case, the region with high conductivity can often be treated in the lumped capacitance model, as a "lump" of material with a simple thermal capacitance consisting of its aggregate heat capacity. Such regions warm or cool, but show no significant temperature *variation* across their extent, during the process (as compared to the rest of the system). This is due to their far higher conductance. During transient conduction, therefore, the temperature across their conductive regions changes uniformly in space, and as a simple exponential in time. An example of such systems are those that follow Newton's law of cooling during transient cooling (or the reverse during heating). The equivalent thermal circuit consists of a simple capacitor in series with a resistor. In such cases, the remainder of the system with high thermal resistance (comparatively low conductivity) plays the role of the resistor in the circuit.

Relativistic Conduction

The theory of relativistic heat conduction is a model that is compatible with the theory of special relativity. For most of the last century, it was recognized that the Fourier equation is in contradiction with the theory of relativity because it admits an infinite speed of propagation of heat signals. For example, according to the Fourier equation, a pulse of heat at the origin would be felt at infinity instantaneously. The speed of information propagation is faster than the speed of light in vacuum, which is physically inadmissible within the framework of relativity.

Quantum Conduction

Second sound is a quantum mechanical phenomenon in which heat transfer occurs by wave-like motion, rather than by the more usual mechanism of diffusion. Heat takes the place of pressure in normal sound waves. This leads to a very high thermal conductivity. It is known as "second sound" because the wave motion of heat is similar to the propagation of sound in air.

Fourier's Law

The law of heat conduction, also known as Fourier's law, states that the rate of heat transfer through a material is proportional to the negative gradient in the temperature and to the area, at right angles to that gradient, through which the heat flows. We can state this law in two equivalent forms: the integral form, in which we look at the amount of energy flowing into or out of a body as a whole, and the differential form, in which we look at the flow rates or fluxes of energy locally.

Newton's law of cooling is a discrete analogue of Fourier's law, while Ohm's law is the electrical analogue of Fourier's law.

Differential Form

The differential form of Fourier's law of thermal conduction shows that the local heat flux density, q, is equal to the product of thermal conductivity, k, and the negative local temperature gradient, $-\nabla T$. The heat flux density is the amount of energy that flows through a unit area per unit time.

$$q = -k\nabla T$$

where (including the SI units)

q is the local heat flux density, $W \cdot m^{-2}$

k is the material's conductivity, $W \cdot m^{-1} \cdot K^{-1}$,

∇T is the temperature gradient, $K \cdot m^{-1}$.

The thermal conductivity, k, is often treated as a constant, though this is not always true. While the thermal conductivity of a material generally varies with temperature, the variation can be small over a significant range of temperatures for some common materials. In anisotropic materials, the thermal conductivity typically varies with orientation; in this case k is represented by a second-order tensor. In non-uniform materials, k varies with spatial location.

For many simple applications, Fourier's law is used in its one-dimensional form. In the x-direction,

$$q_x = -k\frac{dT}{dx}$$

Integral Form

By integrating the differential form over the material's total surface , we arrive at the integral form of Fourier's law:

$$\frac{\partial Q}{\partial t} = -k \oiint_S \nabla T \cdot dS$$

where (including the SI units):

- $\frac{\partial Q}{\partial t}$ is the amount of heat transferred per unit time (in W), and
- dS is an oriented surface area element (in m^2)

The above differential equation, when integrated for a homogeneous material of 1-D geometry between two endpoints at constant temperature, gives the heat flow rate as:

$$\frac{Q}{\Delta t} = -kA\frac{\Delta T}{\Delta x}$$

where

- A is the cross-sectional surface area,

- ΔT is the temperature difference between the ends,
- Δx is the distance between the ends.

This law forms the basis for the derivation of the heat equation.

Conductance

Writing:

$$U = \frac{k}{\Delta x},$$

where U is the conductance, in W/(m² K).

Fourier's law can also be stated as:

$$\frac{\Delta Q}{\Delta t} = UA(-\Delta T).$$

The reciprocal of conductance is resistance, R, given by:

$$R = \frac{1}{U} = \frac{\Delta x}{k} = \frac{A(-\Delta T)}{\frac{\Delta Q}{\Delta t}}.$$

Resistance is additive when several conducting layers lie between the hot and cool regions, because A and Q are the same for all layers. In a multilayer partition, the total conductance is related to the conductance of its layers by:

$$\frac{1}{U} = \frac{1}{U_1} + \frac{1}{U_2} + \frac{1}{U_3} + \cdots$$

So, when dealing with a multilayer partition, the following formula is usually used:

$$\frac{\Delta Q}{\Delta t} = \frac{A(-\Delta T)}{\frac{\Delta x_1}{k_1} + \frac{\Delta x_2}{k_2} + \frac{\Delta x_3}{k_3} + \cdots}.$$

For heat conduction from one fluid to another through a barrier, it is sometimes important to consider the conductance of the thin film of fluid that remains stationary next to the barrier. This thin film of fluid is difficult to quantify because its characteristics depend upon complex conditions of turbulence and viscosity—but when dealing with thin high-conductance barriers it can sometimes be quite significant.

Intensive-property Representation

The previous conductance equations, written in terms of extensive properties, can be reformulated in terms of intensive properties. Ideally, the formulae for conductance should produce a quantity with dimensions independent of distance, like Ohm's Law for electrical resistance, $R = V / I$, and conductance, $G = I / V$.

From the electrical formula: $R = \rho x / A$, where ρ is resistivity, x is length, and A is cross-sectional area, we have $G = kA / x$, where G is conductance, k is conductivity, x is length, and A is cross-sectional area.

For Heat,

$$U = \frac{kA}{\Delta x},$$

where U is the conductance.

Fourier's law can also be stated as:

$$\dot{Q} = U \Delta T,$$

analogous to Ohm's law, $I = V / R$ or $I = VG$.

The reciprocal of conductance is resistance, R, given by:

$$R = \frac{\Delta T}{\dot{Q}},$$

analogous to Ohm's law, $R = V / I$.

The rules for combining resistances and conductances (in series and in parallel) are the same for both heat flow and electric current.

Cylindrical Shells

Conduction through cylindrical shells (e.g. pipes) can be calculated from the internal radius, , the external radius, r_2, the length, ℓ, and the temperature difference between the inner and outer wall, $T_2 - T_1$.

The surface area of the cylinder is $A_r = 2\pi r \ell$

When Fourier's equation is applied:

$$\dot{Q} = -kA_r \frac{dT}{dr} = -2k\pi r \ell \frac{dT}{dr}$$

and rearranged:

$$\dot{Q} \int_{r_1}^{r_2} \frac{1}{r} dr = -2k\pi \ell \int_{T_1}^{T_2} dT$$

then the rate of heat transfer is:

$$\dot{Q} = 2k\pi \ell \frac{T_1 - T_2}{\ln(r_2 / r_1)}$$

the thermal resistance is:

$$R_c = \frac{\Delta T}{\dot{Q}} = \frac{\ln(r_2 / r_1)}{2\pi k \ell}$$

and $\dot{Q} = 2\pi k \ell r_m \frac{T_1 - T_2}{r_2 - r_1}$, where $r_m = \frac{r_2 - r_1}{\ln(r_2 / r_1)}$. It is important to note that this is the log-mean radius.

Spherical

The conduction through a spherical shell with internal radius, r_1, and external radius, r_2, can be calculated in a similar manner as for a cylindrical shell.

The surface area of the sphere is: $A = 4\pi r^2$.

Solving in a similar manner as for a cylindrical shell produces:

$$\dot{Q} = 4k\pi \frac{T_1 - T_2}{1/r_1 - 1/r_2} = 4k\pi \frac{(T_1 - T_2) r_1 r_2}{r_2 - r_1}$$

Transient Thermal Conduction

Interface Heat Transfer

The heat transfer at an interface is considered a transient heat flow. To analyze this problem, the Biot number is important to understand how the system behaves. The Biot number is determined by: $Bi = \frac{hL}{k}$ The heat transfer coefficient h, is introduced in this formula, and is measured in $\frac{J}{m^2 sK}$. If the system has a Biot number of less than 0.1, the material behaves according to Newtonian cooling, i.e. with negligible temperature gradient within the body. If the Biot number is greater than 0.1, the system behaves as a series solution. The temperature profile in terms of time can be derived from the equation:

$$q = -h\Delta T,$$

which becomes:

$$\frac{T - T_f}{T_i - T_f} = \exp\left[\frac{-hAt}{\rho C_p V}\right].$$

The heat transfer coefficient, h, is measured in $\frac{\mathrm{W}}{\mathrm{m^2 K}}$, and represents the transfer of heat at an interface between two materials. This value is different at every interface, and is an important concept in understanding heat flow at an interface.

The series solution can be analyzed with a nomogram. A nomogram has relative temperature as the y coordinate and the Fourier number, which is calculated by:

$$Fo = \frac{\alpha t}{L^2}.$$

The Biot number increases as the Fourier number decreases. There are five steps to determine a temperature profile in terms of time.

1. Calculate the Biot number.
2. Determine which relative depth matters, either x or L.
3. Convert time to the Fourier number.
4. Convert T_i to relative temperature with the boundary conditions.
5. Compared required point to trace specified Biot number on the nomogram.

Thermal Conduction Applications

Splat Cooling

Splat cooling is a method for quenching small droplets of molten materials by rapid contact with a cold surface. The particles undergo a characteristic cooling process, with the heat profile at $t=0$ for initial temperature as the maximum at $x=0$ and $T=0$ at $x=-\infty$ and $x=\infty$, and the heat profile at $t=\infty$ for $-\infty \leq x \leq \infty$ as the boundary conditions. Splat cooling rapidly ends in a steady state temperature, and is similar in form to the Gaussian diffusion equation. The temperature profile, with respect to the position and time of this type of cooling, varies with:

$$T(x,t)-T_i=\frac{T_i \Delta X}{2\sqrt{\pi \alpha t}}\exp\left(-\frac{x^2}{4\alpha t}\right)$$

Splat cooling is a fundamental concept that has been adapted for practical use in the form of thermal spraying. The thermal diffusivity coefficient, represented as α, can be written as $\alpha=\frac{k}{\rho C_p}$. This varies according to the material.

Metal Quenching

Metal quenching is a transient heat transfer process in terms of the time temperature transformation (TTT). It is possible to manipulate the cooling process to adjust the phase of a suitable material. For example, appropriate quenching of steel can convert a desirable proportion of its content of austenite to martensite, creating a very tough product. To achieve this, it is necessary to quench at the "nose" (or eutectic) of the TTT diagram. Since materials differ in their Biot numbers, the time it takes for the material to quench, or the Fourier number, varies in practice. In steel, the quenching temperature range is generally from 600 °C to 200 °C. To control the quenching time and to select suitable quenching media, it is necessary to determine the Fourier number from the desired quenching time, the relative temperature drop, and the relevant Biot number. Usually, the correct figures are read from a standard nomogram. By calculating the heat transfer coefficient from this Biot number, one can find a liquid medium suitable for the application.

Thermal Conduction Instruments

Thermal Conductivity Analyzer

Thermal conduction property of any gas under standard conditions of pressure and temperature is a fixed quantity. This property of a known reference gas or known reference gas mixtures can, therefore, be used for certain sensory applications, such as the thermal conductivity analyzer.

The working of this instrument is by principle based on the Wheatstone bridge containing four filaments whose resistances are matched. Whenever a certain gas is passed over such network of filaments, their resistance changes due to the altered thermal conductivity of the filaments and thereby changing the net voltage output from the Wheatstone Bridge. This voltage output will be correlated with the database to identify the gas sample.

Gas Sensor

The principle of thermal conductivity of gases can also be used to measure the concentration of a gas in a binary mixture of gases.

Working: if the same gas is present around the all the Wheatstone bridge filaments, then the same temperature is maintained in all the filaments and hence same resistances are also maintained; resulting in a balanced Wheatstone bridge. However, If dissimilar gas sample (or gas mixture) is passed over one set of two filaments and the reference gas on the other set of two filaments, then the Wheatstone bridge becomes unbalanced. And the resulting net voltage output of the circuit will be correlated with the database to identify the constituents of the sample gas.

Using this technique many unknown gas samples can be identified by comparing their thermal conductivity with other reference gas of known thermal conductivity. The most commonly used reference gas is nitrogen; as the thermal conductivity of most common gases (except hydrogen and helium) are similar to that of nitrogen.

Convection

Convection is the heat transfer due to the bulk movement of molecules within fluids such as gases and liquids, including molten rock (rheid). Convection includes sub-mechanisms of advection (directional bulk-flow transfer of heat), and diffusion (non-directional transfer of energy or mass particles along a concentration gradient).

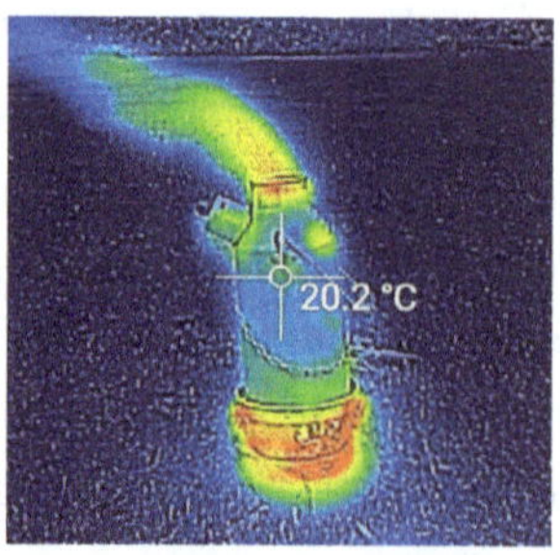

Thermal image of a newly lit Ghillie kettle. The plume of hot air resulting from the convection current is visible.

Convection cannot take place in most solids because neither bulk current flows nor significant diffusion of matter can take place. Diffusion of heat takes place in rigid solids, but that is called heat conduction. Convection, additionally may take place in soft solids or mixtures where solid particles can move past each other.

Thermal convection can be demonstrated by placing a heat source (e.g. a Bunsen burner) at the side of a glass filled with a liquid, and observing the changes in temperature in the glass caused by the warmer fluid circulating into cooler areas.

Convective heat transfer is one of the major types of heat transfer, and convection is also a major mode of mass transfer in fluids. Convective heat and mass transfer takes place both by diffusion – the random Brownian motion of individual particles in the fluid – and by advection, in which matter or heat is transported by the larger-scale motion of currents in the fluid. In the context of heat and mass transfer, the term "convection" is used to refer to the combined effects of advective and diffusive transfer. Sometimes the term "convection" is used to refer specifically to "free heat convection" (natural heat convection) where bulk-flow in a fluid is due to temperature-induced differences in buoyancy, as opposed to "forced heat convection" where forces other than buoyancy (such as pump or fan) move the fluid. However, in mechanics, the correct use of the word "convection" is the more general sense, and different types of convection should be further qualified, for clarity.

Convection can be qualified in terms of being natural, forced, gravitational, granular, or thermomagnetic. It may also be said to be due to combustion, capillary action, or Marangoni and Weissenberg effects. Heat transfer by natural convection plays a role in the structure of Earth's atmosphere, its oceans, and its mantle. Discrete convective cells in the atmosphere can be seen as clouds, with stronger convection resulting in thunderstorms. Natural convection also plays a role in stellar physics.

The convection mechanism is also used in cooking, when using a convection oven, which uses fans to circulate hot air around food in order to cook the food faster than a conventional oven.

Examples and Applications of Convection

Convection occurs on a large scale in atmospheres, oceans, planetary mantles, and it provides the mechanism of heat transfer for a large fraction of the outermost interiors of our sun and all stars. Fluid movement during convection may be invisibly slow, or it may be obvious and rapid, as in a hurricane. On astronomical scales, convection of gas and dust is thought to occur in the accretion disks of black holes, at speeds which may closely approach that of light.

Heat Transfer

A heat sink provides a large surface area for convection to efficiently carry away heat.

Convective heat transfer is a mechanism of heat transfer occurring because of bulk motion (observable movement) of fluids. Heat is the entity of interest being advected (carried), and diffused (dispersed). This can be contrasted with conductive heat transfer, which is the transfer of energy by vibrations at a molecular level through a solid or fluid, and radiative heat transfer, the transfer of energy through electromagnetic waves.

Heat is transferred by convection in numerous examples of naturally occurring fluid flow, such as wind, oceanic currents, and movements within the Earth's mantle. Convection is also used in engineering practices of homes, industrial processes, cooling of equipment, etc.

The rate of convective heat transfer may be improved by the use of a heat sink, often in conjunction with a fan. For instance, a typical computer CPU will have a purpose-made fan to ensure its operating temperature is kept within tolerable limits.

Convection Cells

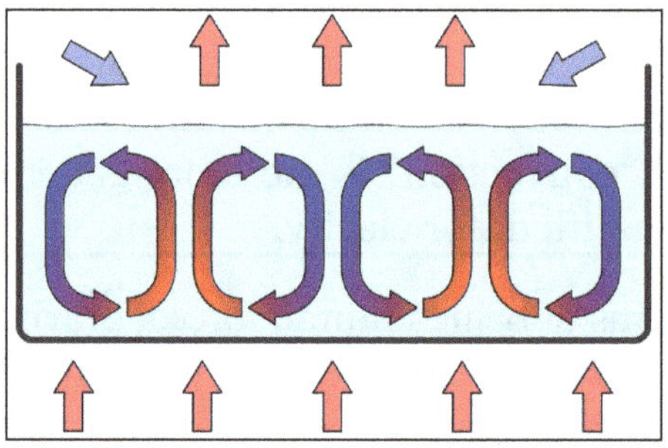

Convection cells in a gravity field.

A convection cell, also known as a Bénard cell is a characteristic fluid flow pattern in many convection systems. A rising body of fluid typically loses heat because it encounters a colder surface. In liquid, this occurs because it exchanges heat with colder liquid through direct exchange. In the example of the Earth's atmosphere, this occurs because it radiates heat. Because of this heat loss the fluid becomes denser than the fluid underneath it, which is still rising. Since it cannot descend through the rising fluid, it moves to one side. At some distance, its downward force overcomes the rising force beneath it, and the fluid begins to descend. As it descends, it warms again and the cycle repeats itself.

Atmospheric Circulation

Atmospheric circulation is the large-scale movement of air, and is a means by which thermal energy is distributed on the surface of the Earth, together with the much slower (lagged) ocean circulation system. The large-scale structure of the atmospheric circulation varies from year to year, but the basic climatological structure remains fairly constant.

Latitudinal circulation occurs because incident solar radiation per unit area is highest at the heat equator, and decreases as the latitude increases, reaching minima at the poles. It consists of two primary convection cells, the Hadley cell and the polar vortex, with the Hadley cell experiencing stronger convection due to the release of latent heat energy by condensation of water vapor at higher altitudes during cloud formation.

Longitudinal circulation, on the other hand, comes about because the ocean has a higher specific heat capacity than land (and also thermal conductivity, allowing the heat to penetrate further beneath the surface) and thereby absorbs and releases more heat, but the temperature changes less than land. This brings the sea breeze, air cooled by the water, ashore in the day, and carries the land breeze, air cooled by contact with the ground, out to sea during the night. Longitudinal circulation consists of two cells, the Walker circulation and El Niño / Southern Oscillation.

Weather

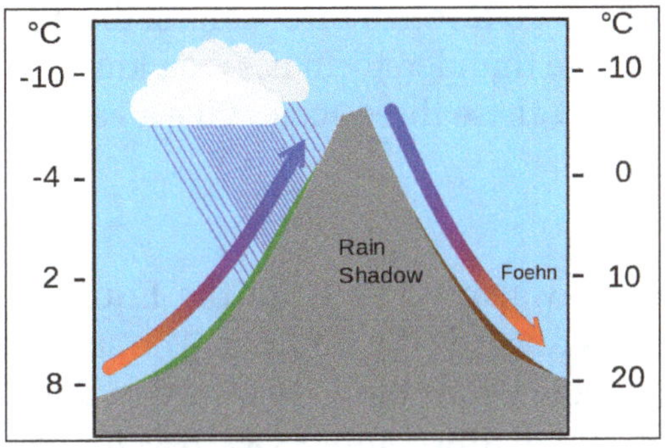

How Foehn is produced.

Some more localized phenomena than global atmospheric movement are also due to convection, including wind and some of the hydrologic cycle. For example, a foehn wind is a down-slope wind which occurs on the downwind side of a mountain range. It results from the adiabatic warming of air which has dropped most of its moisture on windward slopes. Because of the different adiabatic lapse rates of moist and dry air, the air on the leeward slopes becomes warmer than at the same height on the windward slopes.

A thermal column (or thermal) is a vertical section of rising air in the lower altitudes of the Earth's atmosphere. Thermals are created by the uneven heating of the Earth's surface from solar radiation. The Sun warms the ground, which in turn warms the air directly above it. The warmer air expands, becoming less dense than the surrounding air mass, and creating a thermal low. The mass of lighter air rises, and as it does, it cools by expansion at lower air pressures. It stops rising when it has cooled to the same temperature as the surrounding air. Associated with a thermal is a downward flow surrounding the thermal column. The downward moving exterior is caused by colder air being displaced at the top of the thermal. Another convection-driven weather effect is the sea breeze.

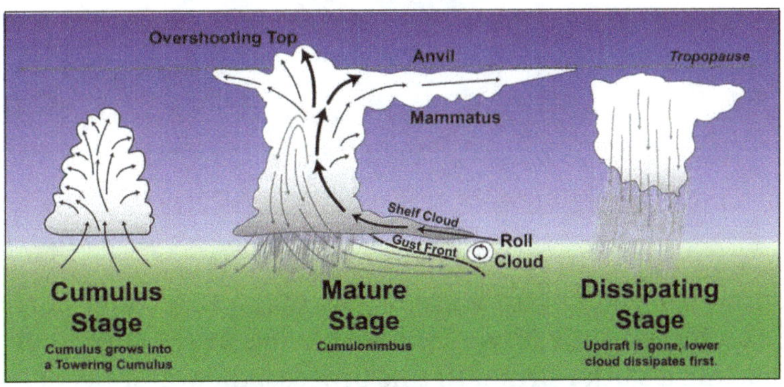

Stages of a thunderstorm's life.

Warm air has a lower density than cool air, so warm air rises within cooler air, similar to hot air balloons. Clouds form as relatively warmer air carrying moisture rises within cooler air. As the moist air rises, it cools, causing some of the water vapor in the rising packet of air to condense. When the moisture condenses, it releases energy known as latent heat of condensation which allows the rising packet of air to cool less than its surrounding air, continuing the cloud's ascension. If enough instability is present in the atmosphere, this process will continue long enough for cumulonimbus clouds to form, which support lightning and thunder. Generally, thunderstorms require three conditions to form: moisture, an unstable airmass, and a lifting force (heat).

All thunderstorms, regardless of type, go through three stages: the developing stage, the mature stage, and the dissipation stage. The average thunderstorm has a 24 km (15 mi) diameter. Depending on the conditions present in the atmosphere, these three stages take an average of 30 minutes to go through.

Oceanic Circulation

Solar radiation affects the oceans: warm water from the Equator tends to circulate toward the poles, while cold polar water heads towards the Equator. The surface currents are initially dictated by surface wind conditions. The trade winds blow westward in the tropics, and the westerlies blow eastward at mid-latitudes. This wind pattern applies a stress to the subtropical ocean surface with negative curl across the Northern Hemisphere, and the reverse across the Southern Hemisphere. The resulting Sverdrup transport is equatorward. Because of conservation of potential vorticity caused by the poleward-moving winds on the subtropical ridge's western periphery and the increased relative vorticity of poleward moving water, transport is balanced by a narrow, accelerating poleward current, which flows along the western boundary of the ocean basin, outweighing the effects of friction with the cold western boundary current which originates from high latitudes. The overall process, known as western intensification, causes currents on the western boundary of an ocean basin to be stronger than those on the eastern boundary.

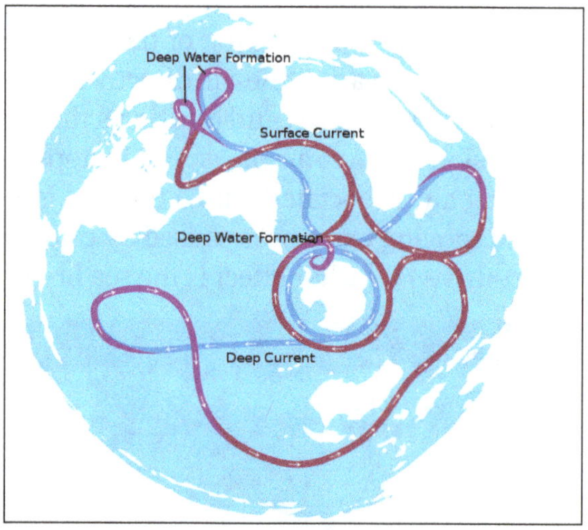

Ocean currents.

As it travels poleward, warm water transported by strong warm water current undergoes evaporative cooling. The cooling is wind driven: wind moving over water cools the water and also causes evaporation, leaving a saltier brine. In this process, the water becomes saltier and denser. and decreases in temperature. Once sea ice forms, salts are left out of the ice, a process known as brine exclusion. These two processes produce water that is denser and colder. The water across the northern Atlantic ocean becomes so dense that it begins to sink down through less salty and less dense water. (The convective action is not unlike that of a lava lamp.) This downdraft of heavy, cold and dense water becomes a part of the North Atlantic Deep Water, a southgoing stream.

Mantle Convection

Mantle convection is the slow creeping motion of Earth's rocky mantle caused by convection currents carrying heat from the interior of the earth to the surface. It is one of 3 driving forces that causes tectonic plates to move around the Earth's surface.

The Earth's surface is divided into a number of tectonic plates that are continuously being created and consumed at their opposite plate boundaries. Creation (accretion) occurs as mantle is added to the growing edges of a plate. This hot added material cools down by conduction and convection of heat. At the consumption edges of the plate, the material has thermally contracted to become dense, and it sinks under its own weight in the process of subduction at an ocean trench. This subducted material sinks to some depth in the Earth's interior where it is prohibited from sinking further. The subducted oceanic crust triggers volcanism.

Stack Effect

The Stack effect or chimney effect is the movement of air into and out of buildings, chimneys, flue gas stacks, or other containers due to buoyancy. Buoyancy occurs due to a difference in indoor-to-outdoor air density resulting from temperature and moisture differences. The greater the thermal difference and the height of the structure, the greater the buoyancy force, and thus the stack effect. The stack effect helps drive natural ventilation and infiltration. Some cooling towers operate on this principle; similarly the solar updraft tower is a proposed device to generate electricity based on the stack effect.

Stellar Physics

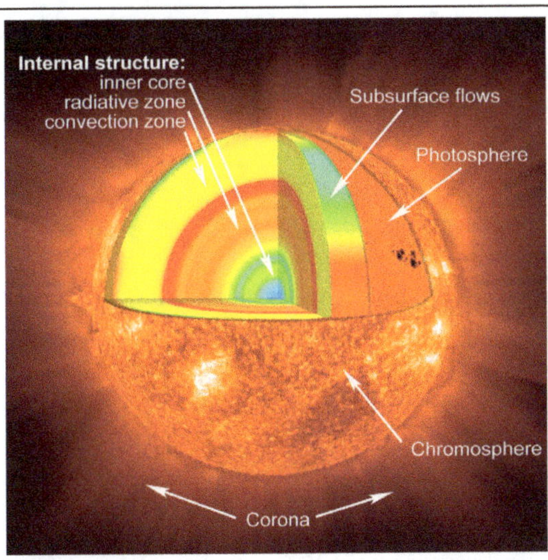

An illustration of the structure of the Sun and a red giant star, showing their convective zones. These are the granular zones in the outer layers of these stars.

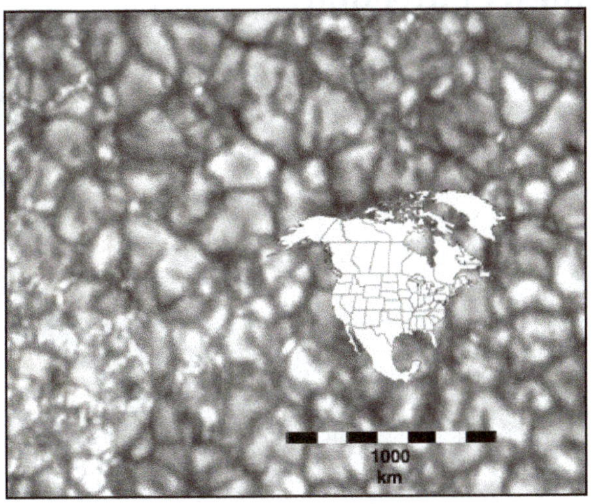

Granules—the tops or upper visible sizes of convection cells, seen on the photosphere of the Sun. These are caused by the convection in the upper photosphere of the Sun.

The convection zone of a star is the range of radii in which energy is transported primarily by convection.

Granules on the photosphere of the Sun are the visible tops of convection cells in the photosphere, caused by convection of plasma in the photosphere. The rising part of the granules is located in the center where the plasma is hotter. The outer edge of the granules is darker due to the cooler descending plasma. A typical granule has a diameter on the order of 1,000 kilometers and each lasts 8 to 20 minutes before dissipating. Below the photosphere is a layer of much larger "super-granules" up to 30,000 kilometers in diameter, with lifespans of up to 24 hours.

Cooking

A convection oven is an oven that has fans to circulate air around food, using the convection mechanism to cook food faster than a conventional oven. Convection ovens distribute heat evenly around the food, removing the blanket of cooler air that surrounds food when it is first placed in an oven and allowing food to cook more evenly in less time and at a lower temperature than in a conventional oven. A convection oven has a fan with a heating element around it. A small fan circulates the air in the cooking chamber.

Convection Mechanisms

Convection may happen in fluids at all scales larger than a few atoms. There are a variety of circumstances in which the forces required for natural and forced convection arise, leading to different types of convection. In broad terms, convection arises because of body forces acting within the fluid, such as gravity.

The causes of convection are generally described as one of either "natural" ("free") or "forced", although other mechanisms also exist. However, the distinction between natural and forced convection is particularly important for convective heat transfer.

Natural Convection

This color schlieren image reveals thermal convection from a human hand (in silhouette) to the surrounding still atmosphere.

Natural convection, or free convection, occurs due to temperature differences which affect the density, and thus relative buoyancy, of the fluid. Heavier (denser) components will fall, while lighter (less dense) components rise, leading to bulk fluid movement. Natural convection can only occur, therefore, in a gravitational field. A common example of natural convection is the rise of smoke from a fire. It can be seen in a pot of boiling water in which the hot and less-dense water on the bottom layer moves upwards in plumes, and the cool and more dense water near the top of the pot likewise sinks.

Natural convection will be more likely and more rapid with a greater variation in density between the two fluids, a larger acceleration due to gravity that drives the convection or a larger distance through the convecting medium. Natural convection will be less likely and less rapid with more rapid diffusion (thereby diffusing away the thermal gradient that is causing the convection) or a more viscous (sticky) fluid.

The onset of natural convection can be determined by the Rayleigh number (Ra).

Note that differences in buoyancy within a fluid can arise for reasons other than temperature variations, in which case the fluid motion is called gravitational convection. However, all types of buoyant convection, including natural convection, do not occur in microgravity environments. All require the presence of an environment which experiences g-force (proper acceleration).

Forced Convection

In forced convection, also called heat advection, fluid movement results from external surface forces such as a fan or pump. Forced convection is typically used to increase the rate of heat exchange. Many types of mixing also utilize forced convection to distribute one substance within another. Forced convection also occurs as a by-product to other processes, such as the action of a propeller in a fluid or aerodynamic heating. Fluid radiator systems, and also heating and cooling of parts of the body by blood circulation, are other familiar examples of forced convection.

Forced convection may happen by natural means, such as when the heat of a fire causes expansion of air and bulk air flow by this means. In microgravity, such flow (which happens in all directions) along with diffusion is the only means by which fires are able to draw in fresh oxygen to maintain themselves. The shock wave that transfers heat and mass out of explosions is also a type of forced convection.

Although forced convection from thermal gas expansion in zero-g does not fuel a fire as well as natural convection in a gravity field, some types of artificial forced convection are far more efficient than free convection, as they are not limited by natural mechanisms. For instance, a convection oven works by forced convection, as a fan which rapidly circulates hot air forces heat into food faster than would naturally happen due to simple heating without the fan.

Gravitational or Buoyant Convection

Gravitational convection is a type of natural convection induced by buoyancy variations resulting from material properties other than temperature. Typically this is caused by a variable composition of the fluid. If the varying property is a concentration gradient, it is known as solutal convection. For example, gravitational convection can be seen in the diffusion of a source of dry salt downward into wet soil due to the buoyancy of fresh water in saline.

Variable salinity in water and variable water content in air masses are frequent causes of convection in the oceans and atmosphere which do not involve heat, or else involve additional compositional density factors other than the density changes from thermal expansion. Similarly, variable composition within the Earth's interior which has not yet achieved maximal stability and minimal energy (in other words, with densest parts deepest) continues to cause a fraction of the convection of fluid rock and molten metal within the Earth's interior.

Gravitational convection, like natural thermal convection, also requires a g-force environment in order to occur.

Granular Convection

Vibration-induced convection occurs in powders and granulated materials in containers subject to vibration where an axis of vibration is parallel to the force of gravity. When the container accelerates upward, the bottom of the container pushes the entire contents upward. In contrast, when the container accelerates downward, the sides of the container push the adjacent material downward by friction, but the material more remote from the sides is less affected. The net result is a slow circulation of particles downward at the sides, and upward in the middle.

If the container contains particles of different sizes, the downward-moving region at the sides is often narrower than the largest particles. Thus, larger particles tend to become sorted to the top of such a mixture. This is one possible explanation of the Brazil nut effect.

Solid-state Convection in Ice

Ice convection on Pluto is believed to occur in a soft mixture of nitrogen ice and carbon monoxide ice. It has also been proposed for Europa, and other bodies in the outer solar system.

Thermomagnetic Convection

Thermomagnetic convection can occur when an external magnetic field is imposed on a ferrofluid with varying magnetic susceptibility. In the presence of a temperature gradient this results in a nonuniform magnetic body force, which leads to fluid movement. A ferrofluid is a liquid which becomes strongly magnetized in the presence of a magnetic field.

This form of heat transfer can be useful for cases where conventional convection fails to provide adequate heat transfer, e.g., in miniature microscale devices or under reduced gravity conditions.

Capillary Action

Capillary action is a phenomenon where liquid spontaneously rises in a narrow space such as a thin tube, or in porous materials. This effect can cause liquids to flow against the force of gravity. It occurs because of inter-molecular attractive forces between the liquid and solid surrounding surfaces; If the diameter of the tube is sufficiently small, then the combination of surface tension and forces of adhesion between the liquid and container act to lift the liquid.

Marangoni Effect

The Marangoni effect is the convection of fluid along an interface between dissimilar substances because of variations in surface tension. Surface tension can vary because of inhomogeneous composition of the substances or the temperature-dependence of surface tension forces. In the latter case the effect is known as thermo-capillary convection.

A well-known phenomenon exhibiting this type of convection is the "tears of wine".

Weissenberg Effect

The Weissenberg effect is a phenomenon that occurs when a spinning rod is placed into a solution of liquid polymer. Entanglements cause the polymer chains to be drawn towards the rod instead of being thrown outward as would happen with an ordinary fluid (i.e., water).

Combustion

In a zero-gravity environment, there can be no buoyancy forces, and thus no natural (free) convection possible, so flames in many circumstances without gravity smother in their own waste gases. However, flames may be maintained with any type of forced convection (breeze); or (in high oxygen environments in "still" gas environments) entirely from the minimal forced convection that occurs as heat-induced *expansion* (not buoyancy) of gases allows for ventilation of the flame, as waste gases move outward and cool, and fresh high-oxygen gas moves in to take up the low pressure zones created when flame-exhaust water condenses.

Mathematical Models of Convection

Mathematically, convection can be described by the convection–diffusion equation, also known as the generic scalar transport equation.

Quantifying Natural versus Forced Convection

In cases of mixed convection (natural and forced occurring together) one would often like to know how much of the convection is due to external constraints, such as the fluid velocity in the pump, and how much is due to natural convection occurring in the system.

The relative magnitudes of the Grashof number and the square of the Reynolds number determine which form of convection dominates. If $Gr/Re^2 \gg 1$, forced convection may be neglected, whereas if $Gr/Re^2 \ll 1$, natural convection may be neglected. If the ratio, known as the Richardson number, is approximately one, then both forced and natural convection need to be taken into account.

Radiation

Radiation is the emission or transmission of energy in the form of waves or particles through space or through a material medium. This includes:

- Electromagnetic radiation, such as radio waves, microwaves, infrared, visible light, ultraviolet, x-rays, and gamma radiation (γ).

- Particle radiation, such as alpha radiation (α), beta radiation (β), and neutron radiation (particles of non-zero rest energy).
- Acoustic radiation, such as ultrasound, sound, and seismic waves (dependent on a physical transmission medium).
- Gravitational radiation, radiation that takes the form of gravitational waves, or ripples in the curvature of spacetime.

Radiation is often categorized as either ionizing or non-ionizing depending on the energy of the radiated particles. Ionizing radiation carries more than 10 eV, which is enough to ionize atoms and molecules, and break chemical bonds. This is an important distinction due to the large difference in harmfulness to living organisms. A common source of ionizing radiation is radioactive materials that emit α, β, or γ radiation, consisting of helium nuclei, electrons or positrons, and photons, respectively. Other sources include X-rays from medical radiography examinations and muons, mesons, positrons, neutrons and other particles that constitute the secondary cosmic rays that are produced after primary cosmic rays interact with Earth's atmosphere.

Gamma rays, X-rays and the higher energy range of ultraviolet light constitute the ionizing part of the electromagnetic spectrum. The word "ionize" refers to the breaking of one or more electrons away from an atom, an action that requires the relatively high energies that these electromagnetic waves supply. Further down the spectrum, the non-ionizing lower energies of the lower ultraviolet spectrum cannot ionize atoms, but can disrupt the inter-atomic bonds which form molecules, thereby breaking down molecules rather than atoms; a good example of this is sunburn caused by long-wavelength solar ultraviolet. The waves of longer wavelength than UV in visible light, infrared and microwave frequencies cannot break bonds but can cause vibrations in the bonds which are sensed as heat. Radio wavelengths and below generally are not regarded as harmful to biological systems. These are not sharp delineations of the energies; there is some overlap in the effects of specific frequencies.

The word radiation arises from the phenomenon of waves *radiating* (i.e., traveling outward in all directions) from a source. This aspect leads to a system of measurements and physical units that are applicable to all types of radiation. Because such radiation expands as it passes through space, and as its energy is conserved (in vacuum), the intensity of all types of radiation from a point source follows an inverse-square law in relation to the distance from its source. Like any ideal law, the inverse-square law approximates a measured radiation intensity to the extent that the source approximates a geometric point.

Ionizing Radiation

Radiation with sufficiently high energy can ionize atoms; that is to say it can knock electrons off atoms, creating ions. Ionization occurs when an electron is stripped (or "knocked out") from an electron shell of the atom, which leaves the atom with a net positive charge. Because living cells and, more importantly, the DNA in those cells can be damaged by this ionization, exposure to ionizing radiation is considered to increase the risk of cancer. Thus "ionizing radiation" is somewhat artificially separated from particle radiation and electromagnetic radiation, simply due to its great potential for biological damage. While an individual cell is made of trillions of atoms, only a small fraction of those will be ionized at low to moderate radiation powers. The probability of ionizing radiation causing cancer is dependent upon the absorbed dose of the radiation, and is a function of the damaging tendency of the type of radiation (equivalent dose) and the sensitivity of the irradiated organism or tissue (effective dose).

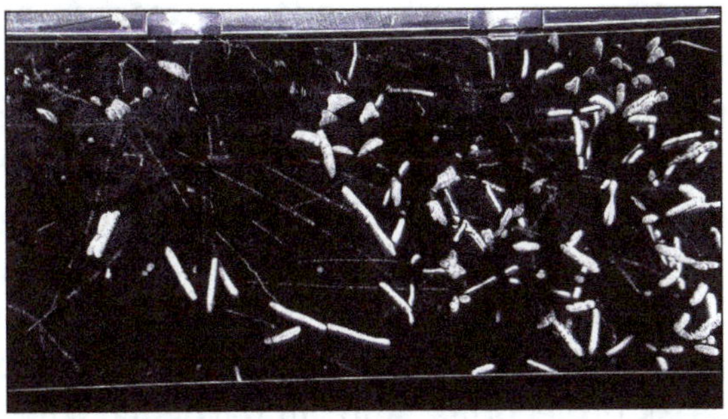

Some kinds of ionizing radiation can be detected in a cloud chamber.

If the source of the ionizing radiation is a radioactive material or a nuclear process such as fission or fusion, there is particle radiation to consider. Particle radiation is subatomic particle accelerated to relativistic speeds by nuclear reactions. Because of their momenta they are quite capable of knocking out electrons and ionizing materials, but since most have an electrical charge, they don't have the penetrating power of ionizing radiation. The exception is neutron particles. There are several different kinds of these particles, but the majority are alpha particles, beta particles, neutrons, and protons. Roughly speaking, photons and particles with energies above about 10 electron volts (eV) are ionizing (some authorities use 33 eV, the ionization energy for water). Particle radiation from radioactive material or cosmic rays almost invariably carries enough energy to be ionizing.

Most ionizing radiation originates from radioactive materials and space (cosmic rays), and as such is naturally present in the environment, since most rocks and soil have small concentrations of radioactive materials. Since this radiation is invisible and not directly detectable by human senses, instruments such as Geiger counters are usually required to detect its presence. In some cases, it may lead to secondary emission of visible light upon its interaction with matter, as in the case of Cherenkov radiation and radio-luminescence.

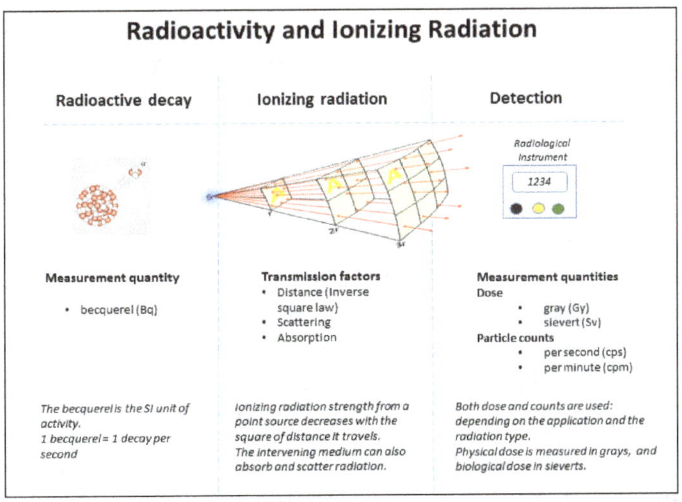

Graphic showing relationships between radioactivity and detected ionizing radiation.

Ionizing radiation has many practical uses in medicine, research and construction, but presents a health hazard if used improperly. Exposure to radiation causes damage to living tissue; high doses result in Acute radiation syndrome (ARS), with skin burns, hair loss, internal organ failure and death, while any dose may result in an increased chance of cancer and genetic damage; a particular form of cancer, thyroid cancer, often occurs when nuclear weapons and reactors are the radiation source because of the biological proclivities of the radioactive iodine fission product, iodine-131. However, calculating the exact risk and chance of cancer forming in cells caused by ionizing radiation is still not well understood and currently estimates are loosely determined by population based data from the atomic bombings of Hiroshima and Nagasaki and from follow-up of reactor accidents, such as the Chernobyl disaster. The International Commission on Radiological Protection states that "The Commission is aware of uncertainties and lack of precision of the models and parameter values", "Collective effective dose is not intended as a tool for epidemiological risk assessment, and it is inappropriate to use it in risk projections" and "in particular, the calculation of the number of cancer deaths based on collective effective doses from trivial individual doses should be avoided."

Ultraviolet Radiation

Ultraviolet, of wavelengths from 10 nm to 125 nm, ionizes air molecules, causing it to be strongly absorbed by air and by ozone (O_3) in particular. Ionizing UV therefore does not penetrate Earth's atmosphere to a significant degree, and is sometimes referred to as vacuum ultraviolet. Although present in space, this part of the UV spectrum is not of biological importance, because it does not reach living organisms on Earth.

There is a zone of the atmosphere in which ozone absorbs some 98% of non-ionizing but dangerous UV-C and UV-B. This so-called ozone layer starts at about 20 miles (32 km) and extends upward. Some of the ultraviolet spectrum that does reach the ground (the part that begins above energies of 3.1 eV, a wavelength less than 400 nm) is non-ionizing, but is still biologically hazardous due to the ability of single photons of this energy to cause electronic excitation in biological molecules, and thus damage them by means of unwanted reactions. An example is the formation of pyrimidine dimers in DNA, which begins at wavelengths below 365 nm (3.4 eV), which is well below ionization energy. This property gives the ultraviolet spectrum some of the dangers of ionizing radiation in biological systems without actual ionization occurring. In contrast, visible light and longer-wavelength electromagnetic radiation, such as infrared, microwaves, and radio waves, consists of photons with too little energy to cause damaging molecular excitation, and thus this radiation is far less hazardous per unit of energy.

X-ray

X-rays are electromagnetic waves with a wavelength less than about 10^{-9} m (greater than 3×10^{17} Hz and 1,240 eV). A smaller wavelength corresponds to a higher energy according to the equation $E=hc/\lambda$. ("E" is Energy; "h" is Planck's constant; "c" is the speed of light; "λ" is wavelength.) When an X-ray photon collides with an atom, the atom may absorb the energy of the photon and boost an electron to a higher orbital level or if the photon is very energetic, it may knock an electron from the atom altogether, causing the atom to ionize. Generally, larger atoms are more likely to absorb an X-ray photon since they have greater energy differences between orbital electrons. Soft tissue in the human body is composed of smaller atoms than the calcium atoms that make up bone, hence there is a contrast in the absorption of X-rays. X-ray machines are specifically designed to take advantage of the absorption difference between bone and soft tissue, allowing physicians to examine structure in the human body.

X-rays are also totally absorbed by the thickness of the earth's atmosphere, resulting in the prevention of the X-ray output of the sun, smaller in quantity than that of UV but nonetheless powerful, from reaching the surface.

Gamma Radiation

Gamma (γ) radiation consists of photons with a wavelength less than 3×10^{-11} meters (greater than 10^{19} Hz and 41.4 keV). Gamma radiation emission is a nuclear process that occurs to rid an unstable nucleus of excess energy after most nuclear reactions. Both alpha and beta particles have an electric charge and mass, and thus are quite likely to interact with other atoms in their path. Gamma radiation, however, is composed of photons, which have neither mass nor electric charge and, as a result, penetrates much further through matter than either alpha or beta radiation.

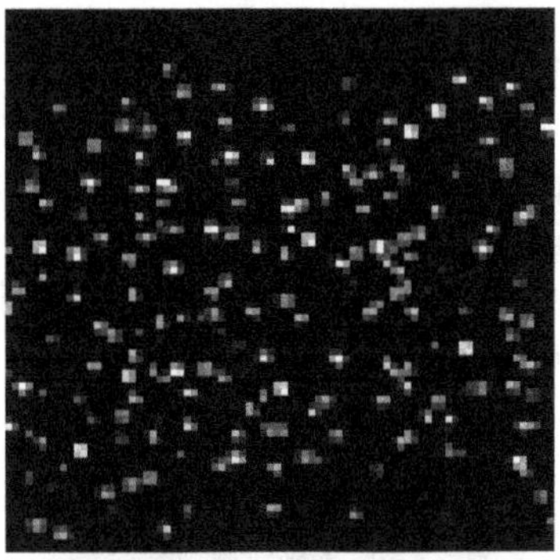

Gamma radiation detected in an isopropanol cloud chamber.

Gamma rays can be stopped by a sufficiently thick or dense layer of material, where the stopping power of the material per given area depends mostly (but not entirely) on the total mass along the path of the radiation, regardless of whether the material is of high or low density. However, as is the case with X-rays, materials with high atomic number such as lead or depleted uranium add a modest (typically 20% to 30%) amount of stopping power over an equal mass of less dense and lower atomic weight materials (such as water or concrete). The atmosphere absorbs all gamma rays approaching Earth from space. Even air is capable of absorbing gamma rays, halving the energy of such waves by passing through, on the average, 500 ft (150 m).

Alpha Radiation

Alpha particles are helium-4 nuclei (two protons and two neutrons). They interact with matter strongly due to their charges and combined mass, and at their usual velocities only penetrate a few centimeters of air, or a few millimeters of low density material (such as the thin mica material which is specially placed in some Geiger counter tubes to allow alpha particles in). This means that alpha particles from ordinary alpha decay do not penetrate the outer layers of dead skin cells and cause no damage to the live tissues below. Some very high energy alpha particles compose about 10% of cosmic rays, and these are capable of penetrating the body and even thin metal plates. However, they are of danger only to astronauts, since they are deflected by the Earth's magnetic field and then stopped by its atmosphere.

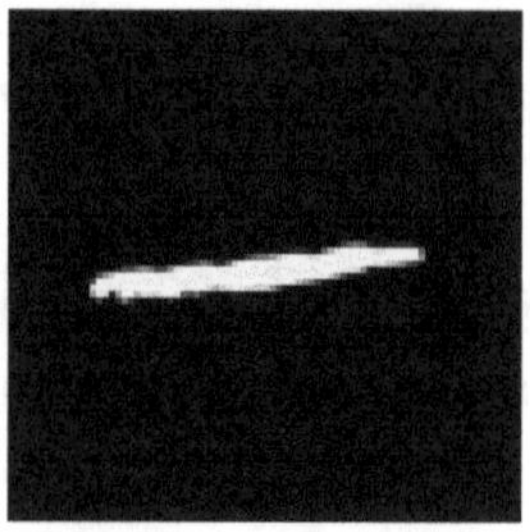

Alpha particle detected in an isopropanol cloud chamber.

Alpha radiation is dangerous when alpha-emitting radioisotopes are ingested or inhaled (breathed or swallowed). This brings the radioisotope close enough to sensitive live tissue for the alpha radiation to damage cells. Per unit of energy, alpha particles are at least 20 times more effective at cell-damage as gamma rays and X-rays. Examples of highly poisonous alpha-emitters are all isotopes of radium, radon, and polonium, due to the amount of decay that occur in these short half-life materials.

Beta Radiation

Beta-minus (β^-) radiation consists of an energetic electron. It is more penetrating than alpha radiation, but less than gamma. Beta radiation from radioactive decay can be stopped with a few centimeters of plastic or a few millimeters of metal. It occurs when a neutron decays into a proton in a nucleus, releasing the beta particle and an antineutrino. Beta radiation from linac accelerators is far more energetic and penetrating than natural beta radiation. It is sometimes used therapeutically in radiotherapy to treat superficial tumors.

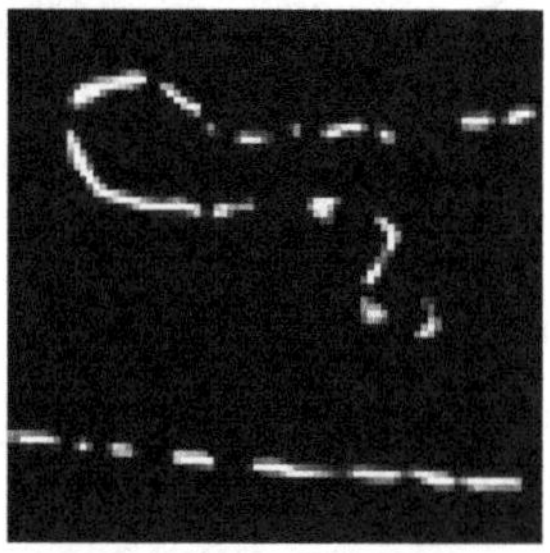

Electrons (beta radiation) detected in an isopropanol cloud chamber.

Beta-plus (β^+) radiation is the emission of positrons, which are the antimatter form of electrons. When a positron slows to speeds similar to those of electrons in the material, the positron will annihilate an electron, releasing two gamma photons of 511 keV in the process. Those two gamma photons will be traveling in (approximately) opposite direction. The gamma radiation from positron annihilation consists of high energy photons, and is also ionizing.

Neutron Radiation

Neutrons are categorized according to their speed/energy. Neutron radiation consists of free neutrons. These neutrons may be emitted during either spontaneous or induced nuclear fission. Neutrons are rare radiation particles; they are produced in large numbers only where chain reaction fission or fusion reactions are active; this happens for about 10 microseconds in a thermonuclear explosion, or continuously inside an operating nuclear reactor; production of the neutrons stops almost immediately in the reactor when it goes non-critical.

Neutrons are the only type of ionizing radiation that can make other objects, or material, radioactive. This process, called neutron activation, is the primary method used to produce radioactive sources for use in medical, academic, and industrial applications. Even comparatively low speed thermal neutrons cause neutron activation (in fact, they cause it more efficiently). Neutrons do not ionize atoms in the same way that charged particles such as protons and electrons do (by the excitation of an electron), because neutrons have no charge. It is through their absorption by nuclei which then become unstable that they cause ionization. Hence, neutrons are said to be "indirectly ionizing." Even neutrons without significant kinetic energy are indirectly ionizing, and are thus a significant radiation hazard. Not all materials are capable of neutron activation; in water, for example, the most common isotopes of both types atoms present (hydrogen and oxygen) capture neutrons and become heavier but remain stable forms of those atoms. Only the absorption of more than one neutron, a statistically rare occurrence, can activate a hydrogen atom, while oxygen requires two additional absorptions. Thus water is only very weakly capable of activation. The sodium in salt (as in sea water), on the other hand, need only absorb a single neutron to become Na-24, a very intense source of beta decay, with half-life of 15 hours.

In addition, high-energy (high-speed) neutrons have the ability to directly ionize atoms. One mechanism by which high energy neutrons ionize atoms is to strike the nucleus of an atom and knock the atom out of a molecule, leaving one or more electrons behind as the chemical bond is broken. This leads to production of chemical free radicals. In addition, very high energy neutrons can cause ionizing radiation by "neutron spallation" or knockout, wherein neutrons cause emission of high-energy protons from atomic nuclei (especially hydrogen nuclei) on impact. The last process imparts most of the neutron's energy to the proton, much like one billiard ball striking another. The charged protons and other products from such reactions are directly ionizing.

High-energy neutrons are very penetrating and can travel great distances in air (hundreds or even thousands of meters) and moderate distances (several meters) in common solids. They typically require hydrogen rich shielding, such as concrete or water, to block them within distances of less than a meter. A common source of neutron radiation occurs inside a nuclear reactor, where a meters-thick water layer is used as effective shielding.

Cosmic Radiation

There are two sources of high energy particles entering the Earth's atmosphere from outer space: the sun and deep space. The sun continuously emits particles, primarily free protons, in the solar wind, and occasionally augments the flow hugely with coronal mass ejections (CME).

The particles from deep space (inter- and extra-galactic) are much less frequent, but of much higher energies. These particles are also mostly protons, with much of the remainder consisting of helions (alpha particles). A few completely ionized nuclei of heavier elements are present. The origin of these galactic cosmic rays is not yet well understood, but they seem to be remnants of supernovae and especially gamma-ray bursts (GRB), which feature magnetic fields capable of the huge accelerations measured from these particles. They may also be generated by quasars, which are galaxy-wide jet phenomena similar to GRBs but known for their much larger size, and which seem to be a violent part of the universe's early history.

Non-ionizing Radiation

The kinetic energy of particles of non-ionizing radiation is too small to produce charged ions when passing through matter. For non-ionizing electromagnetic radiation, the associated particles (photons) have only sufficient energy to change the rotational, vibrational or electronic valence configurations of molecules and atoms. The effect of non-ionizing forms of radiation on living tissue has only recently been studied. Nevertheless, different biological effects are observed for different types of non-ionizing radiation.

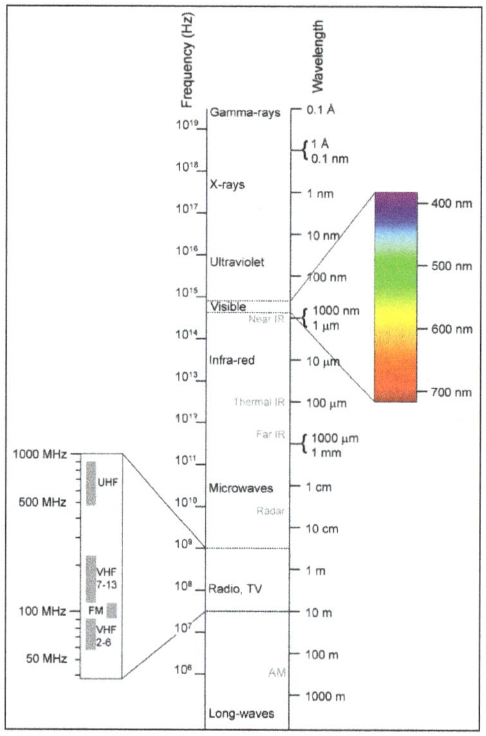

The electromagnetic spectrum.

Even "non-ionizing" radiation is capable of causing thermal-ionization if it deposits enough heat to raise temperatures to ionization energies. These reactions occur at far higher energies than with ionization radiation, which requires only single particles to cause ionization. A familiar example of thermal ionization is the flame-ionization of a common fire, and the browning reactions in common food items induced by infrared radiation, during broiling-type cooking.

The electromagnetic spectrum is the range of all possible electromagnetic radiation frequencies. The electromagnetic spectrum (usually just spectrum) of an object is the characteristic distribution of electromagnetic radiation emitted by, or absorbed by, that particular object.

The non-ionizing portion of electromagnetic radiation consists of electromagnetic waves that are not energetic enough to detach electrons from atoms or molecules and hence cause their ionization. These include radio waves, microwaves, infrared, and (sometimes) visible light. The lower frequencies of ultraviolet light may cause chemical changes and molecular damage similar to ionization, but is technically not ionizing. The highest frequencies of ultraviolet light, as well as all X-rays and gamma-rays are ionizing.

The occurrence of ionization depends on the energy of the individual particles or waves, and not on their number. An intense flood of particles or waves will not cause ionization if these particles or waves do not carry enough energy to be ionizing, unless they raise the temperature of a body to a point high enough to ionize small fractions of atoms or molecules by the process of thermal-ionization (this, however, requires relatively extreme radiation intensities).

Ultraviolet Light

The lower part of the spectrum of ultraviolet, called soft UV, from 3 eV to about 10 eV, is non-ionizing. However, the effects of non-ionizing ultraviolet on chemistry and the damage to biological systems exposed to it (including oxidation, mutation, and cancer) are such that even this part of ultraviolet is often compared with ionizing radiation.

Visible Light

Light, or visible light, is a very narrow range of electromagnetic radiation of a wavelength that is visible to the human eye, or 380–750 nm which equates to a frequency range of 790 to 400 THz respectively. More broadly, physicists use the term "light" to mean electromagnetic radiation of all wavelengths, whether visible or not.

Infrared

Infrared (IR) light is electromagnetic radiation with a wavelength between 0.7 and 300 micrometers, which corresponds to a frequency range between 430 and 1 THz respectively. IR wavelengths are longer than that of visible light, but shorter than that of microwaves. Infrared may be detected at a distance from the radiating objects by "feel." Infrared sensing snakes can detect and focus infrared by use of a pinhole lens in their heads, called "pits". Bright sunlight provides an irradiance of just over 1 kilowatt per square meter at sea level. Of this energy, 53% is infrared radiation, 44% is visible light, and 3% is ultraviolet radiation.

Microwave

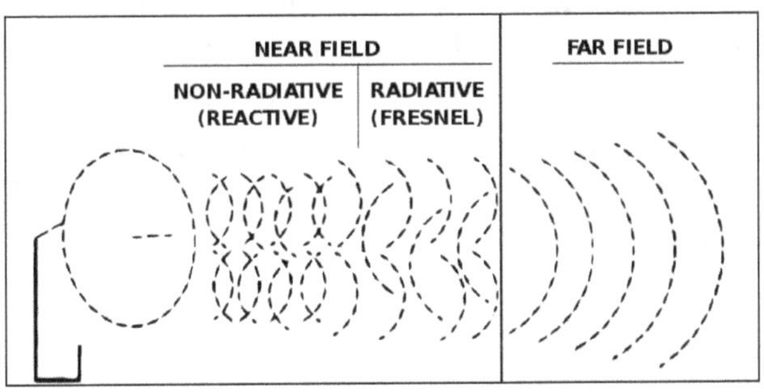

In electromagnetic radiation (such as microwaves from an antenna, shown here) the term "radiation" applies only to the parts of the electromagnetic field that radiate into infinite space and decrease in intensity by an inverse-square law of power so that the total radiation energy that crosses through an imaginary spherical surface is the same, no matter how far away from the antenna the spherical surface is drawn. Electromagnetic radiation includes the far field part of the electromagnetic field around a transmitter. A part of the "near-field" close to the transmitter, is part of the changing electromagnetic field, but does not count as electromagnetic radiation.

Microwaves are electromagnetic waves with wavelengths ranging from as short as one millimeter to as long as one meter, which equates to a frequency range of 300 MHz to 300 GHz. This broad definition includes both UHF and EHF (millimeter waves), but various sources use different other limits. In all cases, microwaves include the entire super high frequency band (3 to 30 GHz, or 10 to 1 cm) at minimum, with RF engineering often putting the lower boundary at 1 GHz (30 cm), and the upper around 100 GHz (3mm).

Radio Waves

Radio waves are a type of electromagnetic radiation with wavelengths in the electromagnetic spectrum longer than infrared light. Like all other electromagnetic waves, they travel at the speed of light. Naturally occurring radio waves are made by lightning, or by certain astronomical objects. Artificially generated radio waves are used for fixed and mobile radio communication, broadcasting, radar and other navigation systems, satellite communication, computer networks and innumerable other applications. In addition, almost any wire carrying alternating current will radiate some of the energy away as radio waves; these are mostly termed interference. Different frequencies of radio waves have different propagation characteristics in the Earth's atmosphere; long waves may bend at the rate of the curvature of the Earth and may cover a part of the Earth very consistently, shorter waves travel around the world by multiple reflections off the ionosphere and the Earth. Much shorter wavelengths bend or reflect very little and travel along the line of sight.

Very Low Frequency

Very low frequency (VLF) refers to a frequency range of 30 Hz to 3 kHz which corresponds to wavelengths of 100,000 to 10,000 meters respectively. Since there is not much bandwidth in this range of the radio spectrum, only the very simplest signals can be transmitted, such as for radio navigation. Also known as the myriameter band or myriameter wave as the wavelengths range from ten to one myriameter (an obsolete metric unit equal to 10 kilometers).

Extremely Low Frequency

Extremely low frequency (ELF) is radiation frequencies from 3 to 30 Hz (10^8 to 10^7 meters respectively). In atmosphere science, an alternative definition is usually given, from 3 Hz to 3 kHz. In the related magnetosphere science, the lower frequency electromagnetic oscillations (pulsations occurring below ~3 Hz) are considered to lie in the ULF range, which is thus also defined differently from the ITU Radio Bands. A massive military ELF antenna in Michigan radiates very slow messages to otherwise unreachable receivers, such as submerged submarines.

Thermal Radiation (Heat)

Thermal radiation is a common synonym for infrared radiation emitted by objects at temperatures often encountered on Earth. Thermal radiation refers not only to the radiation itself, but also the process by which the surface of an object radiates its thermal energy in the form of black body radiation. Infrared or red radiation from a common household radiator or electric heater is an example of thermal radiation, as is the heat emitted by an operating incandescent light bulb. Thermal radiation is generated when energy from the movement of charged particles within atoms is converted to electromagnetic radiation.

Even low-frequency thermal radiation may cause temperature-ionization whenever it deposits sufficient thermal energy to raises temperatures to a high enough level. Common examples of this are the ionization (plasma) seen in common flames, and the molecular changes caused by the "browning" during food-cooking, which is a chemical process that begins with a large component of ionization.

Black-body Radiation

Black-body radiation is an idealized spectrum of radiation emitted by a body that is at a uniform temperature. The shape of the spectrum and the total amount of energy emitted by the body is a function of the absolute temperature of that body. The radiation emitted covers the entire electromagnetic spectrum and the intensity of the radiation (power/unit-area) at a given frequency is described by Planck's law of radiation. For a given temperature of a black-body there is a particular frequency at which the radiation emitted is at its maximum intensity. That maximum radiation frequency moves toward higher frequencies as the temperature of the body increases. The frequency at which the black-body radiation is at maximum is given by Wien's displacement law and is a function of the body's absolute temperature. A black-body is one that emits at any temperature the maximum possible amount of radiation at any given wavelength. A black-body will also absorb the maximum possible incident radiation at any given wavelength. A black-body with a temperature at or below room temperature would thus appear absolutely black, as it would not reflect any incident light nor would it emit enough radiation at visible wavelengths for our eyes to detect. Theoretically, a black-body emits electromagnetic radiation over the entire spectrum from very low frequency radio waves to x-rays, creating a continuum of radiation.

The color of a radiating black-body tells the temperature of its radiating surface. It is responsible for the color of stars, which vary from infrared through red (2,500K), to yellow (5,800K), to white and to blue-white (15,000K) as the peak radiance passes through those points in the visible spectrum. When the peak is below the visible spectrum the body is black, while when it is above the body is blue-white, since all the visible colors are represented from blue decreasing to red.

Uses

Medicine

Radiation and radioactive substances are used for diagnosis, treatment, and research. X-rays, for example, pass through muscles and other soft tissue but are stopped by dense materials. This property of X-rays enables doctors to find broken bones and to locate cancers that might be growing in the body. Doctors also find certain diseases by injecting a radioactive substance and monitoring the radiation given off as the substance moves through the body. Radiation used for cancer treatment is called ionizing radiation because it forms ions in the cells of the tissues it passes through as it dislodges electrons from atoms. This can kill cells or change genes so the cells cannot grow. Other forms of radiation such as radio waves, microwaves, and light waves are called non-ionizing. They don't have as much energy and are not able to ionize cells.

Communication

All modern communication systems use forms of electromagnetic radiation. Variations in the intensity of the radiation represent changes in the sound, pictures, or other information being transmitted. For example, a human voice can be sent as a radio wave or microwave by making the wave vary to corresponding variations in the voice. Musicians have also experimented with gamma sonification, or using nuclear radiation, to produce sound and music.

Science

Researchers use radioactive atoms to determine the age of materials that were once part of a living organism. The age of such materials can be estimated by measuring the amount of radioactive carbon they contain in a process called radiocarbon dating. Similarly, using other radioactive elements, the age of rocks and other geological features (even some man-made objects) can be determined; this is called Radiometric dating. Environmental scientists use radioactive atoms, known as tracer atoms, to identify the pathways taken by pollutants through the environment.

Radiation is used to determine the composition of materials in a process called neutron activation analysis. In this process, scientists bombard a sample of a substance with particles called neutrons. Some of the atoms in the sample absorb neutrons and become radioactive. The scientists can identify the elements in the sample by studying the emitted radiation.

Possible Damage to Health and Environment from Certain Types of Radiation

Ionizing radiation in certain conditions can cause damage to living organisms, causing cancer or genetic damage.

Non-ionizing radiation in certain conditions also can cause damage to living organisms, such as burns. In 2011, the International Agency for Research on Cancer (IARC) of the World Health Organization (WHO) released a statement adding radio frequency electromagnetic fields (including microwave and millimeter waves) to their list of things which are possibly carcinogenic to humans.

Thermodynamics

Thermodynamics is the science of the relationship between heat, work, temperature, and energy. In broad terms, thermodynamics deals with the transfer of energy from one place to another and from one form to another. The key concept is that heat is a form of energy corresponding to a definite amount of mechanical work.

Heat was not formally recognized as a form of energy until about 1798, when Count Rumford (Sir Benjamin Thompson), a British military engineer, noticed that limitless amounts of heat could be generated in the boring of cannon barrels and that the amount of heat generated is proportional to the work done in turning a blunt boring tool. Rumford's observation of the proportionality between heat generated and work done lies at the foundation of thermodynamics. Another pioneer was the French military engineer Sadi Carnot, who introduced the concept of the heat-engine cycle and the principle of reversibility in 1824. Carnot's work concerned the limitations on the maximum amount of work that can be obtained from a steam engine operating with a high-temperature heat transfer as its driving force. Later that century, these ideas were developed by Rudolf Clausius, a German mathematician and physicist, into the first and second laws of thermodynamics, respectively.

The most important laws of thermodynamics are:

- The zeroth law of thermodynamics: When two systems are each in thermal equilibrium with a third system, the first two systems are in thermal equilibrium with each other. This property makes it meaningful to use thermometers as the "third system" and to define a temperature scale.
- The first law of thermodynamics, or the law of conservation of energy: The change in a system's internal energy is equal to the difference between heat added to the system from its surroundings and work done by the system on its surroundings.
- The second law of thermodynamics: Heat does not flow spontaneously from a colder region to a hotter region, or, equivalently, heat at a given temperature cannot be converted entirely into work. Consequently, the entropy of a closed system, or heat energy per unit temperature, increases over time toward some maximum value. Thus, all closed systems tend toward an equilibrium state in which entropy is at a maximum and no energy is available to do useful work. This asymmetry between forward and backward processes gives rise to what is known as the "arrow of time."
- The third law of thermodynamics: The entropy of a perfect crystal of an element in its most stable form tends to zero as the temperature approaches absolute zero. This allows an absolute scale for entropy to be established that, from a statistical point of view, determines the degree of randomness or disorder in a system.

Although thermodynamics developed rapidly during the 19th century in response to the need to optimize the performance of steam engines, the sweeping generality of the laws of thermodynamics makes them applicable to all physical and biological systems. In particular, the laws of thermodynamics give a complete description of all changes in the energy state of any system and its ability to perform useful work on its surroundings.

Classical thermodynamics does not involve the consideration of individual atoms or molecules. Such concerns are the focus of the branch of thermodynamics known as statistical thermodynamics, or statistical mechanics, which expresses macroscopic thermodynamic properties in terms of the behaviour of individual particles and their interactions. It has its roots in the latter part of the 19th century, when atomic and molecular theories of matter began to be generally accepted.

Thermodynamic States

The application of thermodynamic principles begins by defining a system that is in some sense distinct from its surroundings. For example, the system could be a sample of gas inside a cylinder with a movable piston, an entire steam engine, a marathon runner, the planet Earth, a neutron star, a black hole, or even the entire universe. In general, systems are free to exchange heat, work, and other forms of energy with their surroundings.

A system's condition at any given time is called its thermodynamic state. For a gas in a cylinder with a movable piston, the state of the system is identified by the temperature, pressure, and volume of the gas. These properties are characteristic parameters that have definite values at each state and are independent of the way in which the system arrived at that state. In other words, any change in value of a property depends only on the initial and final states of the system, not on the path followed by the system from one state to another. Such properties are called state functions. In contrast, the work done as the piston moves and the gas expands and the heat the gas absorbs from its surroundings depend on the detailed way in which the expansion occurs.

The behaviour of a complex thermodynamic system, such as Earth's atmosphere, can be understood by first applying the principles of states and properties to its component parts—in this case, water, water vapour, and the various gases making up the atmosphere. By isolating samples of material whose states and properties can be controlled and manipulated, properties and their interrelations can be studied as the system changes from state to state.

Thermodynamic Equilibrium

A particularly important concept is thermodynamic equilibrium, in which there is no tendency for the state of a system to change spontaneously. For example, the gas in a cylinder with a movable piston will be at equilibrium if the temperature and pressure inside are uniform and if the restraining force on the piston is just sufficient to keep it from moving. The system can then be made to change to a new state only by an externally imposed change in one of the state functions, such as the temperature by adding heat or the volume by moving the piston. A sequence of one or more such steps connecting different states of the system is called a process. In general, a system is not in equilibrium as it adjusts to an abrupt change in its environment. For example, when a balloon bursts, the compressed gas inside is suddenly far from equilibrium, and it rapidly expands until it reaches a new equilibrium state. However, the same final state could be achieved by placing the same compressed gas in a cylinder with a movable piston and applying a sequence of many small increments in volume (and temperature), with the system being given time to come to equilibrium after each small increment. Such a process is said to be reversible because the system is at (or near) equilibrium at each step along its path, and the direction of change could be reversed at any point. This example illustrates how two different paths can connect the same initial and final states. The first is irreversible (the balloon bursts), and the second is reversible. The concept of reversible processes is something like motion without friction in mechanics. It represents an idealized limiting case that is very useful in discussing the properties of real systems. Many of the results of thermodynamics are derived from the properties of reversible processes.

Temperature

The concept of temperature is fundamental to any discussion of thermodynamics, but its precise definition is not a simple matter. For example, a steel rod feels colder than a wooden rod at room temperature simply because steel is better at conducting heat away from the skin. It is therefore necessary to have an objective way of measuring temperature. In general, when two objects are brought into thermal contact, heat will flow between them until they come into equilibrium with each other. When the flow of heat stops, they are said to be at the same temperature. The zeroth law of thermodynamics formalizes this by asserting that if an object A is in simultaneous thermal equilibrium with two other objects B and C, then B and C will be in thermal equilibrium with each other if brought into thermal contact. Object A can then play the role of a thermometer through some change in its physical properties with temperature, such as its volume or its electrical resistance.

With the definition of equality of temperature in hand, it is possible to establish a temperature scale by assigning numerical values to certain easily reproducible fixed points. For example, in the Celsius (°C) temperature scale, the freezing point of pure water is arbitrarily assigned a temperature of 0 °C and the boiling point of water the value of 100 °C. In the Fahrenheit (°F) temperature scale, these same two points are assigned the values 32 °F and 212 °F, respectively. There are absolute temperature scales related to the second law of thermodynamics. The absolute scale related to the Celsius scale is called the Kelvin (K) scale, and that related to the Fahrenheit scale is called the Rankine (°R) scale. These scales are related by the equations K = °C + 273.15, °R = °F + 459.67, and °R = 1.8 K. Zero in both the Kelvin and Rankine scales is at absolute zero.

Work and Energy

Energy has a precise meaning in physics that does not always correspond to everyday language, and yet a precise definition is somewhat elusive. The term work itself acquired a technical meaning with the advent of Newtonian mechanics. For example, a man pushing on a car may feel that he is doing a lot of work, but no work is actually done unless the car moves. The work done is then the product of the force applied by the man multiplied by the distance through which the car moves. If there is no friction and the surface is level, then the car, once set in motion, will continue rolling indefinitely with constant speed. The rolling car has something that a stationary car does not have—it has kinetic energy of motion equal to the work required to achieve that state of motion. The introduction of the concept of energy in this way is of great value in mechanics because, in the absence of friction, energy is never lost from the system, although it can be converted from one form to another. For example, if a coasting car comes to a hill, it will roll some distance up the hill before coming to a temporary stop. At that moment its kinetic energy of motion has been converted into its potential energy of position, which is equal to the work required to lift the car through the same vertical distance. After coming to a stop, the car will then begin rolling back down the hill until it has completely recovered its kinetic energy of motion at the bottom. In the absence of friction, such systems are said to be conservative because at any given moment the total amount of energy (kinetic plus potential) remains equal to the initial work done to set the system in motion.

As the science of physics expanded to cover an ever-wider range of phenomena, it became necessary to include additional forms of energy in order to keep the total amount of energy constant for all closed systems (or to account for changes in total energy for open systems). For example, if work is done to accelerate charged particles, then some of the resultant energy will be stored in the form of electromagnetic fields and carried away from the system as radiation. In turn the electromagnetic energy can be picked up by a remote receiver (antenna) and converted back into an equivalent amount of work. With his theory of special relativity, Albert Einstein realized that energy (E) can also be stored as mass (m) and converted back into energy, as expressed by his famous equation $E = mc^2$, where c is the velocity of light. All of these systems are said to be conservative in the sense that energy can be freely converted from one form to another without limit. Each fundamental advance of physics into new realms has involved a similar extension to the list of the different forms of energy. In addition to preserving the first law of thermodynamics, also called the law of conservation of energy, each form of energy can be related back to an equivalent amount of work required to set the system into motion.

Thermodynamics encompasses all of these forms of energy, with the further addition of heat to the list of different kinds of energy. However, heat is fundamentally different from the others in that the conversion of work (or other forms of energy) into heat is not completely reversible, even in principle. In the example of the rolling car, some of the work done to set the car in motion is inevitably lost as heat due to friction, and the car eventually comes to a stop on a level surface. Even if all the generated heat were collected and stored in some fashion, it could never be converted entirely back into mechanical energy of motion. This fundamental limitation is expressed quantitatively by the second law of thermodynamics.

The role of friction in degrading the energy of mechanical systems may seem simple and obvious, but the quantitative connection between heat and work, as first discovered by Count Rumford, played a key role in understanding the operation of steam engines in the 19th century and similarly for all energy-conversion processes today.

Total Internal Energy

Although classical thermodynamics deals exclusively with the macroscopic properties of materials—such as temperature, pressure, and volume—thermal energy from the addition of heat can be understood at the microscopic level as an increase in the kinetic energy of motion of the molecules making up a substance. For example, gas molecules have translational kinetic energy that is proportional to the temperature of the gas: the molecules can rotate about their centre of mass, and the constituent atoms can vibrate with respect to each other (like masses connected by springs). Additionally, chemical energy is stored in the bonds holding the molecules together, and weaker long-range interactions between the molecules involve yet more energy. The sum total of all these forms of energy constitutes the total internal energy of the substance in a given thermodynamic state. The total energy of a system includes its internal energy plus any other forms of energy, such as kinetic energy due to motion of the system as a whole (e.g., water flowing through a pipe) and gravitational potential energy due to its elevation.

Open Systems

Thermodynamic Potentials

Most real thermodynamic systems are open systems that exchange heat and work with their environment, rather than the closed systems described thus far. For example, living systems are clearly able to achieve a local reduction in their entropy as they grow and develop; they create structures of greater internal energy (i.e., they lower entropy) out of the nutrients they absorb. This does not represent a violation of the second law of thermodynamics, because a living organism does not constitute a closed system.

In order to simplify the application of the laws of thermodynamics to open systems, parameters with the dimensions of energy, known as thermodynamic potentials, are introduced to describe the system. The resulting formulas are expressed in terms of the Helmholtz free energy F and the Gibbs free energy G, named after the 19th-century German physiologist and physicist Hermann von Helmholtz and the contemporaneous American physicist Josiah Willard Gibbs. The key conceptual step is to separate a system from its heat reservoir. A system is thought of as being held at a constant temperature T by a heat reservoir (i.e., the environment), but the heat reservoir is no longer considered to be part of the system. Recall that the internal energy change (ΔU) of a system is given by:

$$\Delta U = Q - W$$

where Q is the heat absorbed and W is the work done. In general, Q and W separately are not state functions, because they are path-dependent. However, if the path is specified to be any reversible isothermal process, then the heat associated with the maximum work (W_{max}) is $Q_{max} = T\Delta S$. With this substitution the above equation can be rearranged as:

$$-W_{max} = \Delta U - T\Delta S$$

Note that here ΔS is the entropy change just of the system being held at constant temperature, such as a battery. Unlike the case of an isolated system as considered previously, it does not include the entropy change of the heat reservoir (i.e., the surroundings) required to keep the temperature constant. If this additional entropy change of the reservoir were included, the total entropy change would be zero, as in the case of an isolated system. Because the quantities U, T, and S on the right-hand side are all state functions, it follows that $-W_{max}$ must also be a state function. This leads to the definition of the Helmholtz free energy:

$$F = U - TS$$

such that, for any isothermal change of the system,

$$\Delta F = \Delta U - T\Delta S$$

is the negative of the maximum work that can be extracted from the system. The actual work extracted could be smaller than the ideal maximum, or even zero, which implies that $W \le -\Delta F$, with equality applying in the ideal limiting case of a reversible process. When the Helmholtz free energy reaches its minimum value, the system has reached its equilibrium state, and no further work can be extracted from it. Thus, the equilibrium condition of maximum entropy for isolated systems becomes the condition of minimum Helmholtz free energy for open systems held at constant temperature. The one additional precaution required is that work done against the atmosphere be included if the system expands or contracts in the course of the process being considered. Typically, processes are specified as taking place at constant volume and temperature in order that no correction is needed.

Although the Helmholtz free energy is useful in describing processes that take place inside a container with rigid walls, most processes in the real world take place under constant pressure rather than constant volume. For example, chemical reactions in an open test tube—or in the growth of a tomato in a garden—take place under conditions of (nearly) constant atmospheric pressure. It is for the description of these cases that the Gibbs free energy was introduced. As previously established, the quantity:

$$-W_{max} = \Delta U - T\Delta S$$

is a state function equal to the change in the Helmholtz free energy. Suppose that the process being considered involves a large change in volume (ΔV), such as happens when water boils to form steam. The work done by the expanding water vapour as it pushes back the surrounding air at pressure P is PΔV. This is the amount of work that is now split out from Wmax by writing it in the form:

$$W_{max} = W'_{max} + P\Delta V,$$

where W′max is the maximum work that can be extracted from the process taking place at constant temperature T and pressure P, other than the atmospheric work (PΔV). Substituting this partition into the above equation for −Wmax and moving the PΔV term to the right-hand side then yields:

$$-W'_{max} = \Delta U + P\Delta V - T\Delta S.$$

This leads to the definition of the Gibbs free energy:

$$G = U + PV - TS$$

such that, for any isothermal change of the system at constant pressure,

$$\Delta G = \Delta U + P\Delta V - T\Delta S$$

is the negative of the maximum work W′max that can be extracted from the system, other than atmospheric work. As before, the actual work extracted could be smaller than the ideal maximum, or even zero, which implies that $W' \leq -\Delta G$, with equality applying in the ideal limiting case of a reversible process. As with the Helmholtz case, when the Gibbs free energy reaches its minimum value, the system has reached its equilibrium state, and no further work can be extracted from it. Thus, the equilibrium condition becomes the condition of minimum Gibbs free energy for open systems held at constant temperature and pressure, and the direction of spontaneous change is always toward a state of lower free energy for the system (like a ball rolling downhill into a valley). Notice in particular that the entropy can now spontaneously decrease (i.e., TΔS can be negative), provided that this decrease is more than offset by the ΔU + PΔV terms in the definition of ΔG. As further discussed below, a simple example is the spontaneous condensation of steam into water. Although the entropy of water is much less than the entropy of steam, the process occurs spontaneously provided that enough heat energy is taken away from the system to keep the temperature from rising as the steam condenses.

A familiar example of free energy changes is provided by an automobile battery. When the battery is fully charged, its Gibbs free energy is at a maximum, and when it is fully discharged (i.e., dead), its Gibbs free energy is at a minimum. The change between these two states is the maximum amount of electrical work that can be extracted from the battery at constant temperature and pressure. The amount of heat absorbed from the environment in order to keep the temperature of the battery constant (represented by the TΔS term) and any work done against the atmosphere (represented by the PΔV term) are automatically taken into account in the energy balance.

Gibbs Free Energy and Chemical Reactions

All batteries depend on some chemical reaction of the form:

$$\text{Reactants} \rightarrow \text{Products}$$

for the generation of electricity or on the reverse reaction as the battery is recharged. The change in free energy (−ΔG) for a reaction could be determined by measuring directly the amount of electrical work that the battery could do and then using the equation $W_{max} = -\Delta G$. However, the power of thermodynamics is that −ΔG can be calculated without having to build every possible battery and measure its performance. If the Gibbs free energies of the individual substances making up a battery are known, then the total free energies of the reactants can be subtracted from the total free energies of the products in order to find the change in Gibbs free energy for the reaction:

$$\Delta G = G_{\text{products}} - G_{\text{reactants}}$$

Once the free energies are known for a wide variety of substances, the best candidates for actual batteries can be quickly discerned. In fact, a good part of the practice of thermodynamics is concerned with determining the free energies and other thermodynamic properties of individual substances in order that ΔG for reactions can be calculated under different conditions of temperature and pressure.

The term reaction can be interpreted in the broadest possible sense as any transformation of matter from one form to another. In addition to chemical reactions, a reaction could be something as simple as ice (reactants) turning to liquid water (products), the nuclear reactions taking place in the interior of stars, or elementary particle reactions in the early universe. No matter what the process, the direction of spontaneous change (at constant temperature and pressure) is always in the direction of decreasing free energy.

Enthalpy and the Heat of Reaction

The free energy change $W_{max} = -\Delta G$ corresponds to the maximum possible useful work that can be extracted from a reaction, such as in an electrochemical battery. This represents one extreme limit of a continuous range of possibilities. At the other extreme, for example, battery terminals can be connected directly by a wire and the reaction allowed to proceed freely without doing any useful work. In this case $W' = 0$, and the first law of thermodynamics for the reaction becomes:

$$\Delta U = Q_0 - P\Delta V$$

where Q_0 is the heat absorbed when the reaction does no useful work and, as before, PΔV is the atmospheric work term. The key point is that the quantities ΔU and PΔV are exactly the same as in the other limiting case, in which the reaction does maximum work. This follows because these quantities are state functions, which depend only on the initial and final states of a system and not on any path connecting the states. The amount of useful work done just represents different paths connecting the same initial and final states. This leads to the definition of enthalpy (H), or heat content, as:

$$H = U + PV.$$

Its significance is that, for a reaction occurring freely (i.e., doing no useful work) at constant temperature and pressure, the heat absorbed is:

$$Q_0 = \Delta U + P\Delta V = \Delta H,$$

where ΔH is called the heat of reaction. The heat of reaction is easy to measure because it simply represents the amount of heat that is given off if the reactants are mixed together in a beaker and allowed to react freely without doing any useful work.

The above definition for enthalpy and its physical significance allow the equation for ΔG to be written in the particularly illuminating and instructive form:

$$\Delta G = \Delta H - T\Delta S.$$

Both terms on the right-hand side represent heats of reaction but under different sets of circumstances. ΔH is the heat of reaction (i.e., the amount of heat absorbed from the surroundings in order to hold the temperature constant) when the reaction does no useful work, and TΔS is the heat of reaction when the reaction does maximum useful work in an electrochemical cell. The (negative) difference between these two heats is exactly the maximum useful work –ΔG that can be extracted from the reaction. Thus, useful work can be obtained by contriving for a system to extract additional heat from the environment and convert it into work. The difference ΔH – TΔS represents the fundamental limitation imposed by the second law of thermodynamics on how much additional heat can be extracted from the environment and converted into useful work for a given reaction mechanism. An electrochemical cell (such as a car battery) is a contrivance by means of which a reaction can be made to do the maximum possible work against an opposing electromotive force, and hence the reaction literally becomes reversible in the sense that a slight increase in the opposing voltage will cause the direction of the reaction to reverse and the cell to start charging up instead of discharging.

As a simple example, consider a reaction in which water turns reversibly into steam by boiling. To make the reaction reversible, suppose that the mixture of water and steam is contained in a cylinder with a movable piston and held at the boiling point of 373 K (100 °C) at 1 atmosphere pressure by a heat reservoir. The enthalpy change is ΔH = 40.65 kilojoules per mole, which is the latent heat of vaporization. The entropy change is:

$$\Delta S = \frac{40.65}{373} = 0.109 \textit{ kilojoules per mole} \cdot K,$$

representing the higher degree of disorder when water evaporates and turns to steam. The Gibbs free energy change is ΔG = ΔH – TΔS. In this case the Gibbs free energy change is zero because the water and steam are in equilibrium, and no useful work can be extracted from the system (other than work done against the atmosphere). In other words, the Gibbs free energy per molecule of water (also called the chemical potential) is the same for both liquid water and steam, and so water molecules can pass freely from one phase to the other with no change in the total free energy of the system.

Thermodynamic Properties and Relations

In order to carry through a program of finding the changes in the various thermodynamic functions that accompany reactions—such as entropy, enthalpy, and free energy—it is often useful to know these quantities separately for each of the materials entering into the reaction. For example, if the entropies are known separately for the reactants and products, then the entropy change for the reaction is just the difference:

$$\Delta S_{\text{reaction}} = S_{\text{products}} - S_{\text{reactants}}$$

and similarly for the other thermodynamic functions. Furthermore, if the entropy change for a reaction is known under one set of conditions of temperature and pressure, it can be found under other sets of conditions by including the variation of entropy for the reactants and products with temperature or pressure as part of the overall process. For these reasons, scientists and engineers have developed extensive tables of thermodynamic properties for many common substances, together with their rates of change with state variables such as temperature and pressure.

The science of thermodynamics provides a rich variety of formulas and techniques that allow the maximum possible amount of information to be extracted from a limited number of laboratory measurements of the properties of materials. However, as the thermodynamic state of a system depends on several variables—such as temperature, pressure, and volume—in practice it is necessary first to decide how many of these are independent and then to specify what variables are allowed to change while others are held constant. For this reason, the mathematical language of partial differential equations is indispensable to the further elucidation of the subject of thermodynamics.

Of especially critical importance in the application of thermodynamics are the amounts of work required to make substances expand or contract and the amounts of heat required to change the temperature of substances. The first is determined by the equation of state of the substance and the second by its heat capacity. Once these physical properties have been fully characterized, they can be used to calculate other thermodynamic properties, such as the free energy of the substance under various conditions of temperature and pressure.

In what follows, it will often be necessary to consider infinitesimal changes in the parameters specifying the state of a system. The first law of thermodynamics then assumes the differential form $dU = d'Q - d'W$. Because U is a state function, the infinitesimal quantity dU must be an exact differential, which means that its definite integral depends only on the initial and final states of the system. In contrast, the quantities $d'Q$ and $d'W$ are not exact differentials, because their integrals can be evaluated only if the path connecting the initial and final states is specified. The examples to follow will illustrate these rather abstract concepts.

Work of Expansion and Contraction

The first task in carrying out the above program is to calculate the amount of work done by a single pure substance when it expands at constant temperature. Unlike the case of a chemical reaction, where the volume can change at constant temperature and pressure because of the liberation of gas, the volume of a single pure substance placed in a cylinder cannot change unless either the pressure or the temperature changes. To calculate the work, suppose that a piston moves by an infinitesimal amount dx. Because pressure is force per unit area, the total restraining force exerted by the piston on the gas is PA, where A is the cross-sectional area of the piston. Thus, the incremental amount of work done is $d'W = PA\,dx$.

However, A dx can also be identified as the incremental change in the volume (dV) swept out by the head of the piston as it moves. The result is the basic equation $d'W = P\,dV$ for the incremental work done by a gas when it expands. For a finite change from an initial volume Vi to a final volume V_f, the total work done is given by the integral:

$$W = \int_{V_i}^{V_f} P\,dV.$$

Because P in general changes as the volume V changes, this integral cannot be calculated until P is specified as a function of V; in other words, the path for the process must be specified. This gives precise meaning to the concept that dW is not an exact differential.

Equations of State

The equation of state for a substance provides the additional information required to calculate the amount of work that the substance does in making a transition from one equilibrium state to another along some specified path. The equation of state is expressed as a functional relationship connecting the various parameters needed to specify the state of the system. The basic concepts apply to all thermodynamic systems, but here, in order to make the discussion specific, a simple gas inside a cylinder with a movable piston will be considered. The equation of state then takes the form of an equation relating P, V, and T, such that if any two are specified, the third is determined. In the limit of low pressures and high temperatures, where the molecules of the gas move almost independently of one another, all gases obey an equation of state known as the ideal gas law: PV = nRT, where n is the number of moles of the gas and R is the universal gas constant, 8.3145 joules per K. In the International System of Units, energy is measured in joules, volume in cubic metres (m^3), force in newtons (N), and pressure in pascals (Pa), where 1 Pa = 1 N/m^2. A force of one newton moving through a distance of one metre does one joule of work. Thus, both the products PV and RT have the dimensions of work (energy).

To illustrate the path-dependence of the work done, consider three processes connecting the same initial and final states. The temperature is the same for both states, but, in going from state i to state f, the gas expands from V_i to V_f (doing work), and the pressure falls from P_i to P_f. According to the definition of the integral in equation $W = \int_{V_t}^{V_f} P\,dV$, the work done is the area under the curve (or straight line) for each of the three processes. For processes I and III the areas are rectangles, and so the work done is:

$$W_I = P_i(V_f - V_i)$$

and

$$W_{III} = P_f(V_f - V_i),$$

respectively. Process II is more complicated because P changes continuously as V changes. However, T remains constant, and so one can use the equation of state to substitute P = nRT/V in equation $W = \int_{V_t}^{V_f} P\,dV$ to obtain:

$$W_{II} = \int_{V_s}^{V_f} \frac{nRT}{V} dV = nRT \ln \frac{V_f}{V_t}$$

or, because:

$$P_i V_i = nRT = P_f V_f$$

for an (ideal gas) isothermal process,

$$W_{II} = P_i V_i \ln \frac{V_f}{V_i} = P_f V_f \ln \frac{V_f}{V_i}.$$

W_{II} is thus the work done in the reversible isothermal expansion of an ideal gas. The amount of work is clearly different in each of the three cases. For a cyclic process the net work done equals the area enclosed by the complete cycle.

Heat Capacity and Specific Heat

As shown originally by Count Rumford, there is an equivalence between heat (measured in calories) and mechanical work (measured in joules) with a definite conversion factor between the two. The conversion factor, known as the mechanical equivalent of heat, is 1 calorie = 4.184 joules. (There are several slightly different definitions in use for the calorie. The calorie used by nutritionists is actually a kilocalorie.) In order to have a consistent set of units, both heat and work will be expressed in the same units of joules.

The amount of heat that a substance absorbs is connected to its temperature change via its molar specific heat c, defined to be the amount of heat required to change the temperature of 1 mole of the substance by 1 K. In other words, c is the constant of proportionality relating the heat absorbed (d′Q) to the temperature change (dT) according to d′Q = nc dT, where n is the number of moles. For example, it takes approximately 1 calorie of heat to increase the temperature of 1 gram of water by 1 K. Since there are 18 grams of water in 1 mole, the molar heat capacity of water is 18 calories per K, or about 75 joules per K. The total heat capacity C for n moles is defined by C = nc.

However, since d′Q is not an exact differential, the heat absorbed is path-dependent and the path must be specified, especially for gases where the thermal expansion is significant. Two common ways of specifying the path are either the constant-pressure path or the constant-volume path. The two different kinds of specific heat are called c_P and c_V respectively, where the subscript denotes the quantity that is being held constant. It should not be surprising that cP is always greater than cV, because the substance must do work against the surrounding atmosphere as it expands upon heating at constant pressure but not at constant volume. In fact, this difference was used by the 19th-century German physicist Julius Robert von Mayer to estimate the mechanical equivalent of heat.

Heat Capacity and Internal Energy

The goal in defining heat capacity is to relate changes in the internal energy to measured changes in the variables that characterize the states of the system. For a system consisting of a single pure substance, the only kind of work it can do is atmospheric work, and so the first law reduces to:

$$dU = d'Q - P\,dV.$$

Suppose now that U is regarded as being a function U(T, V) of the independent pair of variables T and V. The differential quantity dU can always be expanded in terms of its partial derivatives according to:

$$dU = \left(\frac{\partial U}{\partial T}\right)_V dT + \left(\frac{\partial U}{\partial T}\right)_T dV$$

where the subscripts denote the quantity being held constant when calculating derivatives. Substituting this equation into dU = d′Q – P dV then yields the general expression:

$$d'Q = \left(\frac{\partial U}{\partial T}\right)_V dT + \left[P + \left(\frac{\partial U}{\partial T}\right)_T\right] dV$$

for the path-dependent heat. The path can now be specified in terms of the independent variables T and V. For a temperature change at constant volume, dV = 0 and, by definition of heat capacity,

$$d'Q_V = C_V\, dT.$$

The above equation then gives immediately:

$$C_V = \left(\frac{\partial U}{\partial T}\right)_V$$

for the heat capacity at constant volume, showing that the change in internal energy at constant volume is due entirely to the heat absorbed.

To find a corresponding expression for CP, one need only change the independent variables to T and P and substitute the expansion:

$$dV = \left(\frac{\partial V}{\partial T}\right)_P dT + \left(\frac{\partial V}{\partial T}\right)_T dP$$

for dV in equation $dU = d'Q - P\,dV$. and correspondingly for dU to obtain:

$$d'Q = \left[\left(\frac{\partial U}{\partial T}\right)_P + P\left(\frac{\partial V}{\partial T}\right)_P\right] dT + \left[\left(\frac{\partial U}{\partial P}\right)_r + p\left(\frac{\partial V}{\partial P}\right)_r\right] dP.$$

For a temperature change at constant pressure, dP = 0, and, by definition of heat capacity, d′Q = C_p dT, resulting in:

$$C_p = C_v + \left[p + \left(\frac{\partial V}{\partial T}\right)_r\right]\left(\frac{\partial V}{\partial T}\right)_p.$$

The two additional terms beyond C_V have a direct physical meaning. The term:

$$p\left(\frac{\partial V}{\partial T}\right)_P$$

represents the additional atmospheric work that the system does as it undergoes thermal expansion at constant pressure, and the second term involving:

$$\left(\frac{\partial U}{\partial V}\right)_T$$

represents the internal work that must be done to pull the system apart against the forces of attraction between the molecules of the substance (internal stickiness). Because there is no internal stickiness for an ideal gas, this term is zero, and, from the ideal gas law, the remaining partial derivative is:

$$p\left(\frac{\partial V}{\partial T}\right)_P = nR$$

With these substitutions the equation for C_P becomes simply:

$$C_P = C_V + nR$$

Or

$$C_P = C_V + R$$

for the molar specific heats. For example, for a monatomic ideal gas (such as helium), $C_V = 3R/2$ and $C_P = 5R/2$ to a good approximation. $C_V T$ represents the amount of translational kinetic energy possessed by the atoms of an ideal gas as they bounce around randomly inside their container. Diatomic molecules (such as oxygen) and polyatomic molecules (such as water) have additional rotational motions that also store thermal energy in their kinetic energy of rotation. Each additional degree of freedom contributes an additional amount R to C_V. Because diatomic molecules can rotate about two axes and polyatomic molecules can rotate about three axes, the values of C_V increase to 5R/2 and 3R respectively, and C_P correspondingly increases to 7R/2 and 4R. (C_V and C_P increase still further at high temperatures because of vibrational degrees of freedom.) For a real gas such as water vapour, these values are only approximate, but they give the correct order of magnitude. For example, the correct values are $C_P = 37.468$ joules per K (i.e., 4.5R) and $C_P - C_V = 9.443$ joules per K (i.e., 1.14R) for water vapour at 100 °C and 1 atmosphere pressure.

Entropy as an Exact Differential

Because the quantity $dS = d'Q_{max}/T$ is an exact differential, many other important relationships connecting the thermodynamic properties of substances can be derived. For example, with the substitutions $d'Q = T\,dS$ and $d'W = P\,dV$, the differential form ($dU = d'Q - d'W$) of the first law of thermodynamics becomes (for a single pure substance):

$$dU = T\,dS - P\,dV.$$

The advantage gained by the above formula is that dU is now expressed entirely in terms of state functions in place of the path-dependent quantities d'Q and d'W. This change has the very important mathematical implication that the appropriate independent variables are S and V in place of T and V, respectively, for internal energy.

This replacement of T by S as the most appropriate independent variable for the internal energy of substances is the single most valuable insight provided by the combined first and second laws of thermodynamics. With U regarded as a function U(S, V), its differential dU is:

$$dU = \left(\frac{\partial U}{\partial S}\right)_V dS + \left(\frac{\partial U}{\partial V}\right)_S dV.$$

A comparison with the preceding equation shows immediately that the partial derivatives are:

$$\left(\frac{\partial U}{\partial S}\right)_V = T \text{ and } \left(\frac{\partial U}{\partial V}\right)_S = -P.$$

Furthermore, the cross partial derivatives,

$$\left(\frac{\partial^2 U}{\partial V \partial S}\right) = \left(\frac{\partial T}{\partial V}\right)_S \text{ and } \left(\frac{\partial^2 U}{\partial S \partial V}\right) = -\left(\frac{\partial P}{\partial S}\right)_V.$$

must be equal because the order of differentiation in calculating the second derivatives of U does not matter. Equating the right-hand sides of the above pair of equations then yields:

$$\left(\frac{\partial T}{\partial V}\right)_S = -\left(\frac{\partial P}{\partial S}\right)_V.$$

This is one of four Maxwell relations. They are all extremely useful in that the quantity on the right-hand side is virtually impossible to measure directly, while the quantity on the left-hand side is easily measured in the laboratory. For the present case one simply measures the adiabatic variation of temperature with volume in an insulated cylinder so that there is no heat flow (constant S).

The other three Maxwell relations follow by similarly considering the differential expressions for the thermodynamic potentials F(T, V), H(S, P), and G(T, P), with independent variables as indicated. The results are:

$$\left(\frac{\partial P}{\partial T}\right)_V = \left(\frac{\partial S}{\partial V}\right)_T \left(\frac{\partial V}{\partial S}\right)_P = \left(\frac{\partial T}{\partial P}\right)_S \text{ and } \left(\frac{\partial V}{\partial T}\right)_P = -\left(\frac{\partial S}{\partial P}\right)_T.$$

As an example of the use of these equations, equation $C_p = C_v + \left[p + \left(\frac{\partial V}{\partial T}\right)_r\right]\left(\frac{\partial V}{\partial T}\right)_p$ for $C_P - C_V$ contains the partial derivative:

$$\left(\frac{\partial U}{\partial V}\right)_r$$

which vanishes for an ideal gas and is difficult to evaluate directly from experimental data for real substances. The general properties of partial derivatives can first be used to write it in the form:

$$\left(\frac{\partial U}{\partial V}\right)_T = \left(\frac{\partial U}{\partial V}\right)_S + \left(\frac{\partial U}{\partial S}\right)_V \left(\frac{\partial S}{\partial V}\right)_T$$

Combining this with equation $\left(\frac{\partial U}{\partial S}\right)_V = T$ and $\left(\frac{\partial U}{\partial V}\right)_S = -p$ for the partial derivatives together with the first of the Maxwell equations from equation:

$$\left(\frac{\partial P}{\partial T}\right)_V = \left(\frac{\partial S}{\partial V}\right)_T \left(\frac{\partial V}{\partial S}\right)_P = \left(\frac{\partial T}{\partial P}\right)_S \text{ and } \left(\frac{\partial V}{\partial T}\right)_P = -\left(\frac{\partial S}{\partial P}\right)_T.$$

then yields the desired result:

$$\left(\frac{\partial U}{\partial V}\right)_r = -P + T\left(\frac{\partial P}{\partial T}\right)_V$$

The quantity:

$$\left(\frac{\partial P}{\partial T}\right)_V$$

comes directly from differentiating the equation of state. For an ideal gas:

$$\left(\frac{\partial P}{\partial T}\right)_V = \frac{nR}{V} = \frac{P}{T},$$

and so:

$$\left(\frac{\partial U}{\partial V}\right)_T$$

is zero as expected. The departure of:

$$\left(\frac{\partial U}{\partial V}\right)_T$$

from zero reveals directly the effects of internal forces between the molecules of the substance and the work that must be done against them as the substance expands at constant temperature.

The Clausius-clapeyron Equation

Phase changes, such as the conversion of liquid water to steam, provide an important example of a system in which there is a large change in internal energy with volume at constant temperature. Suppose that the cylinder contains both water and steam in equilibrium with each other at pressure P, and the cylinder is held at constant temperature T. The pressure remains equal to the vapour pressure P_{vap} as the piston moves up, as long as both phases remain present. All that happens is that more water turns to steam, and the heat reservoir must supply the latent heat of vaporization, λ = 40.65 kilojoules per mole, in order to keep the temperature constant.

The results of the preceding topic can be applied now to find the variation of the boiling point of water with pressure. Suppose that as the piston moves up, 1 mole of water turns to steam. The change in volume inside the cylinder is then $\Delta V = V_{gas} - V_{liquid}$, where V_{gas} = 30.143 litres is the volume of 1 mole of steam at 100 °C, and V_{liquid} = 0.0188 litre is the volume of 1 mole of water. By the first law of thermodynamics, the change in internal energy ΔU for the finite process at constant P and T is $\Delta U = \lambda - P\Delta V$.

The variation of U with volume at constant T for the complete system of water plus steam is thus:

$$\left(\frac{\partial U}{\partial V}\right)_r = \frac{\Delta U}{\Delta V} = \frac{\lambda}{\Delta V} - P.$$

A comparison with equation $\left(\frac{\partial U}{\partial V}\right)_r = -P + T\left(\frac{\partial P}{\partial T}\right)_V$ then yields the equation:

$$\left(\frac{\partial P}{\partial T}\right)_V = \frac{\lambda}{\Delta V}.$$

However, for the present problem, P is the vapour pressure P_{vapour}, which depends only on T and is independent of V. The partial derivative is then identical to the total derivative:

$$\frac{dP_{vapour}}{dT},$$

giving the Clausius-Clapeyron equation:

$$\left(\frac{\partial P}{\partial T}\right)_V = \frac{\lambda}{\Delta V}.$$

This equation is very useful because it gives the variation with temperature of the pressure at which water and steam are in equilibrium—i.e., the boiling temperature. An approximate but even more useful version of it can be obtained by neglecting V_{liquid} in comparison with V_{ga}s and using:

$$V_{gas} = \frac{RT}{P_{vapour}},$$

from the ideal gas law. The resulting differential equation can be integrated to give:

$$\frac{1}{T} = \frac{1}{T_0} + \frac{R}{\lambda}\ln\frac{P_0}{P}.$$

For example, at the top of Mount Everest, atmospheric pressure is about 30 percent of its value at sea level. Using the values R = 8.3145 joules per K and λ = 40.65 kilojoules per mole, the above equation gives T = 342 K (69 °C) for the boiling temperature of water, which is barely enough to make tea.

Thermodynamic Systems

A thermodynamic system is a group of material and/or radiative contents. Its properties may be described by thermodynamic state variables such as temperature, entropy, internal energy, and pressure.

The simplest state of a thermodynamic system is a state of thermodynamic equilibrium, as opposed to a non-equilibrium state. A system is defined as quantity of matter or a region in space chosen for study. Everything external to the system is surrounding. Thermodynamic system and surrounding is always separated by the boundary.

The system can be separated from its surrounding by a wall or without a wall.

When the state of its content varies in space, the system can be considered as many systems located next to each other, each being a different thermodynamical system.

A thermodynamic system is subject to external interventions called thermodynamic operations; these alter the system's walls or its surroundings; as a result, the system undergoes thermodynamic processes according to the principles of thermodynamics.

The thermodynamic state of a thermodynamic system is its internal state as specified by its state variables. In addition to the state variables, a thermodynamic account also requires a special kind of quantity called a state function, which is a function of the defining state variables. For example, if the state variables are internal energy, volume and mole amounts, that special function is the entropy. These quantities are inter-related by one or more functional relationships called equations of state, and by the system's characteristic equation. Thermodynamics imposes restrictions on the possible equations of state and on the characteristic equation. The restrictions are imposed by the laws of thermodynamics.

According to the permeabilities of the walls of a system, transfers of energy and matter occur between it and its surroundings, which are assumed to be unchanging over time, until a state of thermodynamic equilibrium is attained. The only states considered in equilibrium thermodynamics are equilibrium states. Classical thermodynamics includes equilibrium thermodynamics. It also considers: (a) systems considered in terms of cyclic sequences of processes rather than of states of the system; such were historically important in the conceptual development of the subject; and (b) systems considered in terms of processes described by steady flows; such are important in engineering.

In 1824 Sadi Carnot described a thermodynamic system as the working substance (such as the volume of steam) of any heat engine under study. The very existence of such thermodynamic systems may be considered a fundamental postulate of equilibrium thermodynamics, though it is only rarely cited as a numbered law. According to Bailyn, the commonly rehearsed statement of the zeroth law of thermodynamics is a consequence of this fundamental postulate.

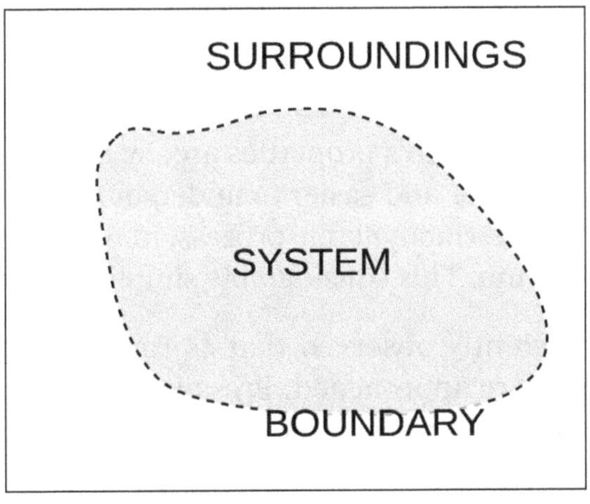

In equilibrium thermodynamics the state variables do not include fluxes because in a state of thermodynamic equilibrium all fluxes have zero values by postulation. Equilibrium thermodynamic processes may involve fluxes but these must have ceased by the time a thermodynamic process or operation is complete bringing a system to its eventual thermodynamic state. Non-equilibrium thermodynamics allows its state variables to include non-zero fluxes, that describe transfers of mass or energy or entropy between a system and its surroundings.

Thermodynamic equilibrium is characterized by absence of flow of mass or energy. Equilibrium thermodynamics, as a subject in physics, considers macroscopic bodies of matter and energy in states of internal thermodynamic equilibrium. It uses the concept of thermodynamic processes, by which bodies pass from one equilibrium state to another by transfer of matter and energy between them. The term 'thermodynamic system' is used to refer to bodies of matter and energy in the special context of thermodynamics. The possible equilibria between bodies are determined by the physical properties of the walls that separate the bodies. Equilibrium thermodynamics in general does not measure time. Equilibrium thermodynamics is a relatively simple and well settled subject. One reason for this is the existence of a well defined physical quantity called 'the entropy of a body'.

Non-equilibrium thermodynamics, as a subject in physics, considers bodies of matter and energy that are not in states of internal thermodynamic equilibrium, but are usually participating in processes of transfer that are slow enough to allow description in terms of quantities that are closely related to thermodynamic state variables. It is characterized by presence of flows of matter and energy. For this topic, very often the bodies considered have smooth spatial inhomogeneities, so that spatial gradients, for example a temperature gradient, are well enough defined. Thus the description of non-equilibrium thermodynamic systems is a field theory, more complicated than the theory of equilibrium thermodynamics. Non-equilibrium thermodynamics is a growing subject, not an established edifice. In general, it is not possible to find an exactly defined entropy for non-equilibrium problems. For many non-equilibrium thermodynamical problems, an approximately defined quantity called 'time rate of entropy production' is very useful.

Another kind of thermodynamic system is considered in engineering. It takes part in a flow process. The account is in terms that approximate, well enough in practice in many cases, equilibrium thermodynamical concepts.

Systems in Equilibrium

At thermodynamic equilibrium, a system's properties are, by definition, unchanging in time. Systems in equilibrium are much simpler and easier to understand than systems not in equilibrium. In some cases, when analyzing a thermodynamic process, one can assume that each intermediate state in the process is at equilibrium. This considerably simplifies the analysis.

In isolated systems it is consistently observed that as time goes on internal rearrangements diminish and stable conditions are approached. Pressures and temperatures tend to equalize, and matter arranges itself into one or a few relatively homogeneous phases. A system in which all processes of change have gone practically to completion is considered in a state of thermodynamic equilibrium. The thermodynamic properties of a system in equilibrium are unchanging in time. Equilibrium system states are much easier to describe in a deterministic manner than non-equilibrium states.

For a process to be reversible, each step in the process must be reversible. For a step in a process to be reversible, the system must be in equilibrium throughout the step. That ideal cannot be accomplished in practice because no step can be taken without perturbing the system from equilibrium, but the ideal can be approached by making changes slowly.

Walls

A system is enclosed by walls that bound it and connect it to its surroundings. Often a wall restricts passage across it by some form of matter or energy, making the connection indirect. Sometimes a wall is no more than an imaginary two-dimensional closed surface through which the connection to the surroundings is direct.

A wall can be fixed (e.g. a constant volume reactor) or moveable (e.g. a piston). For example, in a reciprocating engine, a fixed wall means the piston is locked at its position; then, a constant volume process may occur. In that same engine, a piston may be unlocked and allowed to move in and out. Ideally, a wall may be declared adiabatic, diathermal, impermeable, permeable, or semi-permeable. Actual physical materials that provide walls with such idealized properties are not always readily available.

Types of transfers permitted by types of wall			
Type of wall	Type of transfer		
	Matter	Work	Heat
Permeable to matter	✓	✗	✗
Permeable to energy but impermeable to matter	✗	✓	✓
Adiabatic	✗	✓	✗
Adynamic and impermeable to matter	✗	✗	✓
Isolating	✗	✗	✗

The system is delimited by walls or boundaries, either actual or notional, across which conserved (such as matter and energy) or unconserved (such as entropy) quantities can pass into and out of the system. The space outside the thermodynamic system is known as the *surroundings*, a *reservoir*, or the *environment*. The properties of the walls determine what transfers can occur. A wall that allows transfer of a quantity is said to be permeable to it, and a thermodynamic system is classified by the permeabilities of its several walls. A transfer between system and surroundings can arise by contact, such as conduction of heat, or by long-range forces such as an electric field in the surroundings.

A system with walls that prevent all transfers is said to be isolated. This is an idealized conception, because in practice some transfer is always possible, for example by gravitational forces. It is an axiom of thermodynamics that an isolated system eventually reaches internal thermodynamic equilibrium, when its state no longer changes with time.

The walls of a closed system allow transfer of energy as heat and as work, but not of matter, between it and its surroundings. The walls of an *open system* allow transfer both of matter and of energy. This scheme of definition of terms is not uniformly used, though it is convenient for some purposes. In particular, some writers use 'closed system' where 'isolated system' is here used.

Anything that passes across the boundary and effects a change in the contents of the system must be accounted for in an appropriate balance equation. The volume can be the region surrounding a single atom resonating energy, such as Max Planck defined in 1900; it can be a body of steam or air in a steam engine, such as Sadi Carnot defined in 1824. It could also be just one nuclide (i.e. a system of quarks) as hypothesized in quantum thermodynamics.

Surroundings

The system is the part of the universe being studied, while the *surroundings* is the remainder of the universe that lies outside the boundaries of the system. It is also known as the *environment*, and the *reservoir*. Depending on the type of system, it may interact with the system by exchanging mass, energy (including heat and work), momentum, electric charge, or other conserved properties. The environment is ignored in analysis of the system, except in regards to these interactions.

Closed System

In a closed system, no mass may be transferred in or out of the system boundaries. The system always contains the same amount of matter, but heat and work can be exchanged across the boundary of the system. Whether a system can exchange heat, work, or both is dependent on the property of its boundary.

- Adiabatic boundary – not allowing any heat exchange: A thermally isolated system.
- Rigid boundary – not allowing exchange of work: A mechanically isolated system.

One example is fluid being compressed by a piston in a cylinder. Another example of a closed system is a bomb calorimeter, a type of constant-volume calorimeter used in measuring the heat of combustion of a particular reaction. Electrical energy travels across the boundary to produce a spark between the electrodes and initiates combustion. Heat transfer occurs across the boundary after combustion but no mass transfer takes place either way.

Beginning with the first law of thermodynamics for an open system, this is expressed as:

$$\Delta U = Q - W + m_i(h + \frac{1}{2}v^2 + gz)_i - m_e(h + \frac{1}{2}v^2 + gz)_e$$

where U is internal energy, Q is the heat added to the system, W is the work done by the system, and since no mass is transferred in or out of the system, both expressions involving mass flow are zero and the first law of thermodynamics for a closed system is derived. The first law of thermodynamics for a closed system states that the increase of internal energy of the system equals the amount of heat added to the system minus the work done by the system. For infinitesimal changes the first law for closed systems is stated by:

$$\mathrm{d}U = \delta Q - \delta W.$$

If the work is due to a volume expansion by $\mathrm{d}V$ at a pressure P then:

$$\delta W = P\mathrm{d}V.$$

For a homogeneous system undergoing a reversible process, the second law of thermodynamics reads:

$$\delta Q = T\mathrm{d}S$$

where T is the absolute temperature and S is the entropy of the system. With these relations the fundamental thermodynamic relation, used to compute changes in internal energy, is expressed as:

$$\mathrm{d}U = T\mathrm{d}S - P\mathrm{d}V.$$

For a simple system, with only one type of particle (atom or molecule), a closed system amounts to a constant number of particles. However, for systems undergoing a chemical reaction, there may be all sorts of molecules being generated and destroyed by the reaction process. In this case, the fact that the system is closed is expressed by stating that the total number of each elemental atom is conserved, no matter what kind of molecule it may be a part of. Mathematically:

$$\sum_{j=1}^{m} a_{ij} N_j = b_i^0$$

where N_j is the number of j-type molecules, a_{ij} is the number of atoms of element i in molecule j and b_i^o is the total number of atoms of element i in the system, which remains constant, since the system is closed. There is one such equation for each element in the system.

Isolated System

An isolated system is more restrictive than a closed system as it does not interact with its surroundings in any way. Mass and energy remains constant within the system, and no energy or mass transfer takes place across the boundary. As time passes in an isolated system, internal differences in the system tend to even out and pressures and temperatures tend to equalize, as do density differences. A system in which all equalizing processes have gone practically to completion is in a state of thermodynamic equilibrium.

Truly isolated physical systems do not exist in reality (except perhaps for the universe as a whole), because, for example, there is always gravity between a system with mass and masses elsewhere. However, real systems may behave nearly as an isolated system for finite (possibly very long) times. The concept of an isolated system can serve as a useful model approximating many real-world situations. It is an acceptable idealization used in constructing mathematical models of certain natural phenomena.

In the attempt to justify the postulate of entropy increase in the second law of thermodynamics, Boltzmann's H-theorem used equations, which assumed that a system (for example, a gas) was isolated. That is all the mechanical degrees of freedom could be specified, treating the walls simply as mirror boundary conditions. This inevitably led to Loschmidt's paradox. However, if the stochastic behavior of the molecules in actual walls is considered, along with the randomizing effect of the ambient, background thermal radiation, Boltzmann's assumption of molecular chaos can be justified.

The second law of thermodynamics for isolated systems states that the entropy of an isolated system not in equilibrium tends to increase over time, approaching maximum value at equilibrium. Overall, in an isolated system, the internal energy is constant and the entropy can never decrease. A *closed* system's entropy can decrease e.g. when heat is extracted from the system.

It is important to note that isolated systems are not equivalent to closed systems. Closed systems cannot exchange matter with the surroundings, but can exchange energy. Isolated systems can exchange neither matter nor energy with their surroundings, and as such are only theoretical and do not exist in reality (except, possibly, the entire universe).

It is worth noting that 'closed system' is often used in thermodynamics discussions when 'isolated system' would be correct - i.e. there is an assumption that energy does not enter or leave the system.

Selective Transfer of Matter

For a thermodynamic process, the precise physical properties of the walls and surroundings of the system are important, because they determine the possible processes.

An open system has one or several walls that allow transfer of matter. To account for the internal energy of the open system, this requires energy transfer terms in addition to those for heat and work. It also leads to the idea of the chemical potential.

A wall selectively permeable only to a pure substance can put the system in diffusive contact with a reservoir of that pure substance in the surroundings. Then a process is possible in which that pure substance is transferred between system and surroundings. Also, across that wall a contact equilibrium with respect to that substance is possible. By suitable thermodynamic operations, the pure substance reservoir can be dealt with as a closed system. Its internal energy and its entropy can be determined as functions of its temperature, pressure, and mole number.

A thermodynamic operation can render impermeable to matter all system walls other than the contact equilibrium wall for that substance. This allows the definition of an intensive state variable, with respect to a reference state of the surroundings, for that substance. The intensive variable is called the chemical potential; for component substance i it is usually denoted μ_i. The corresponding extensive variable can be the number of moles N_i of the component substance in the system.

For a contact equilibrium across a wall permeable to a substance, the chemical potentials of the substance must be same on either side of the wall. This is part of the nature of thermodynamic equilibrium, and may be regarded as related to the zeroth law of thermodynamics.

Open System

In an open system, there is an exchange of energy and matter between the system and the surroundings.The presence of reactants in an open beaker is an example of an open system.Here the boundaray is an imaginary surface enclosing the beaker and reactants. It is named *closed*, if borders are impenetrable for substance, but allow transit of energy in the form of heat, and *isolated*, if there is no exchange of heat and substances. The open system cannot exist in the equilibrium state. To describe deviation of the thermodynamic system from equilibrium, in addition to constitutive variables that was described above, a set of internal variables $\xi_1, \xi_2, \ldots$ that are called *internal variables* have been introduced. The equilibrium state is considered to be stable. and the main property of the internal variables, as measures of non-equilibrium of the system, is their trending to disappear; the local law of disappearing can be written as relaxation equation for each internal variable:

$$\frac{d\xi_i}{dt} = -\frac{1}{\tau_i}\left(\xi_i - \xi_i^{(0)}\right), \quad i = 1,2,\ldots,$$

where $\tau_i = \tau_i(T, x_1, x_2, \ldots, x_n)$ is a relaxation time of a corresponding variables. It is convenient to consider the initial value ξ_i^0 are equal to zero.

The essential contribution to the thermodynamics of the open non-equilibrium systems was brought by Prigogine, when he and his collaborators investigated the systems of chemically reacting substances. The stationary states of such systems exists due to exchange both particles and energy with the environment. Prigogine has specified three contributions to the variation of entropy of the considered open system at the given volume and constant temperature T. The increment of entropy S can be calculated according to the formula:

$$T\,dS = \Delta Q - \sum_j \Xi_j \Delta\xi_j + \sum_{\alpha=1}^{k} \mu_\alpha \Delta N_\alpha .$$

The first term on the right hand side of the equation presents a stream of thermal energy into the system; the last term - a stream of energy into the system coming with the stream of particles of substances ΔN_α that can be positive or negative, μ_α is chemical potential of substance α. The middle term in previous equation depicts energy dissipation (entropy production) due to the relaxation of internal variables ξ_j. In the case of chemically reacting substances, which was investigated by Prigogine, the internal variables appear to be measures of incompleteness of chemical reactions, that is measures of how much the considered system with chemical reactions is out of equilibrium. The theory can be generalized, to consider any deviation from the equilibrium state as an internal variable, so that we consider the set of internal variables ξ_j in previous equation to consist of the quantities defining not only degrees of completeness of all chemical reactions occurring in the system, but also the structure of the system, gradients of temperature, difference of concentrations of substances and so on.

The Prigogine approach to the open system allow describing the growth and development of living odjects in thermodynamic terms.

Adiabatic System

An adiabatic system is the one which doesn't allow any heat to be transferred into or out of the system. The $PV^\gamma = constant$ equation is only valid for an adiabatic system which is also undergoing a reversible process provided it is a closed system having an ideal gas. If it fails to satisfy any of these conditions then only $dQ = 0$ is true and it cannot be represented in an equation like $PV^\gamma = constant$.

Laws of Thermodynamics

Zeroth Law of Thermodynamics

The zeroth law of thermodynamics states that if two thermodynamic systems are each in thermal equilibrium with a third one, then they are in thermal equilibrium with each other. Accordingly, thermal equilibrium between systems is a transitive relation.

Two systems are said to be in the relation of thermal equilibrium if they are linked by a wall permeable only to heat and they do not change over time. As a convenience of language, systems are sometimes also said to be in a relation of thermal equilibrium if they are not linked so as to be able to transfer heat to each other, but would still not do so (even) if they were connected by a wall permeable only to heat.

The physical meaning is expressed by Maxwell in the words: "All heat is of the same kind". Another statement of the law is "All diathermal walls are equivalent".

The law is important for the mathematical formulation of thermodynamics, which needs the assertion that the relation of thermal equilibrium is an equivalence relation. This information is needed for a mathematical definition of temperature that will agree with the physical existence of valid thermometers.

Zeroth Law as Equivalence Relation

A thermodynamic system is by definition in its own state of internal thermodynamic equilibrium, that is to say, there is no change in its observable state (i.e. macrostate) over time and no flows occur in it. One precise statement of the zeroth law is that the relation of thermal equilibrium is an equivalence relation on pairs of thermodynamic systems. In other words, the set of all systems each in its own state of internal thermodynamic equilibrium may be divided into subsets in which every system belongs to one and only one subset, and is in thermal equilibrium with every other member of that subset, and is not in thermal equilibrium with a member of any other subset. This means that a unique "tag" can be assigned to every system, and if the "tags" of two systems are the same, they are in thermal equilibrium with each other, and if different, they are not. This property is used to justify the use of empirical temperature as a tagging system. Empirical temperature provides further relations of thermally equilibrated systems, such as order and continuity with regard to "hotness" or "coldness", but these are not implied by the standard statement of the zeroth law.

If it is defined that a thermodynamic system is in thermal equilibrium with itself (i.e., thermal equilibrium is reflexive), then the zeroth law may be stated as follows:

> "If a body C, be in thermal equilibrium with two other bodies, A and B, then A and B are in thermal equilibrium with one another."

This statement asserts that thermal equilibrium is a left-Euclidean relation between thermodynamic systems. If we also define that every thermodynamic system is in thermal equilibrium with itself, then thermal equilibrium is also a reflexive relation. Binary relations that are both reflexive and Euclidean are equivalence relations. Thus, again implicitly assuming reflexivity, the zeroth law is therefore often expressed as a right-Euclidean statement:

> "If two systems are in thermal equilibrium with a third system, then they are in thermal equilibrium with each other."

One consequence of an equivalence relationship is that the equilibrium relationship is symmetric: If *A* is in thermal equilibrium with *B*, then *B* is in thermal equilibrium with *A*. Thus we may say that two systems are in thermal equilibrium with each other, or that they are in mutual equilibrium. Another consequence of equivalence is that thermal equilibrium is a transitive relationship and is occasionally expressed as such:

> "If A is in thermal equilibrium with B and if B is in thermal equilibrium with C, then A is in thermal equilibrium with C ."

A reflexive, transitive relationship does not guarantee an equivalence relationship. In order for the above statement to be true, *both* reflexivity *and* symmetry must be implicitly assumed.

It is the Euclidean relationships which apply directly to thermometry. An ideal thermometer is a thermometer which does not measurably change the state of the system it is measuring. Assuming that the unchanging reading of an ideal thermometer is a valid "tagging" system for the equivalence classes of a set of equilibrated thermodynamic systems, then if a thermometer gives the same reading for two systems, those two systems are in thermal equilibrium, and if we thermally connect the two systems, there will be no subsequent change in the state of either one. If the readings are different, then thermally connecting the two systems will cause a change in the states of both systems and when the change is complete, they will both yield the same thermometer reading. The zeroth law provides no information regarding this final reading.

Foundation of Temperature

The zeroth law establishes thermal equilibrium as an equivalence relationship. An equivalence relationship on a set (such as the set of all systems each in its own state of internal thermodynamic equilibrium) divides that set into a collection of distinct subsets ("disjoint subsets") where any member of the set is a member of one and only one such subset. In the case of the zeroth law, these subsets consist of systems which are in mutual equilibrium. This partitioning allows any member of the subset to be uniquely "tagged" with a label identifying the subset to which it belongs. Although the labeling may be quite arbitrary, temperature is just such a labeling process which uses the real number system for tagging. The zeroth law justifies the use of suitable thermodynamic systems as thermometers to provide such a labeling, which yield any number of possible empirical temperature scales, and justifies the use of the second law of thermodynamics to provide an absolute, or thermodynamic temperature scale. Such temperature scales bring additional continuity and ordering (i.e., "hot" and "cold") properties to the concept of temperature.

In the space of thermodynamic parameters, zones of constant temperature form a surface, that provides a natural order of nearby surfaces. One may therefore construct a global temperature function that provides a continuous ordering of states. The dimensionality of a surface of constant temperature is one less than the number of thermodynamic parameters, thus, for an ideal gas described with three thermodynamic parameters *P*, *V* and *N*, it is a two-dimensional surface.

For example, if two systems of ideal gases are in equilibrium, then $P_1V_1/N_1 = P_2V_2/N_2$ where P_i is the pressure in the *i*th system, V_i is the volume, and N_i is the amount (in moles, or simply the number of atoms) of gas.

The surface PV/N = constant defines surfaces of equal thermodynamic temperature, and one may label defining *T* so that $PV/N = RT$, where *R* is some constant. These systems can now be used as a thermometer to calibrate other systems. Such systems are known as "ideal gas thermometers".

In a sense, focused on in the zeroth law, there is only one kind of diathermal wall or one kind of heat, as expressed by Maxwell's dictum that "All heat is of the same kind". But in another sense, heat is transferred in different ranks, as expressed by Sommerfeld's dictum "Thermodynamics investigates the conditions that govern the transformation of heat into work. It teaches us to recognize temperature as the measure of the work-value of heat. Heat of higher temperature is richer, is capable of doing more work. Work may be regarded as heat of an infinitely high temperature, as unconditionally available heat." This is why temperature is the particular variable indicated by the zeroth law's statement of equivalence.

Physical Meaning of the Zeroth Law

In Carathéodory's theory, it is postulated that there exist walls "permeable only to heat", though heat is not explicitly defined in that paper. This postulate is a physical postulate of existence. It does not, however, as worded just previously, say that there is only one kind of heat. This paper of Carathéodory states as proviso 4 of its account of such walls: "Whenever each of the systems S_1 and S_2 is made to reach equilibrium with a third system S_3 under identical conditions, systems S_1 and S_2 are in mutual equilibrium". It is the function of this statement in the paper, not there labeled as the zeroth law, to provide not only for the existence of transfer of energy other than by work or transfer of matter, but further to provide that such transfer is unique in the sense that there is only one kind of such wall, and one kind of such transfer. This is signaled in the postulate of this paper of Carathéodory that precisely one non-deformation variable is needed to complete the specification of a thermodynamic state, beyond the necessary deformation variables, which are not restricted in number. It is therefore not exactly clear what Carathéodory means when in the introduction of this paper he writes: "It is possible to develop the whole theory without assuming the existence of heat, that is of a quantity that is of a different nature from the normal mechanical quantities."

Maxwell discusses at some length ideas which he summarizes by the words "All heat is of the same kind". Modern theorists sometimes express this idea by postulating the existence of a unique one-dimensional *hotness manifold*, into which every proper temperature scale has a monotonic mapping. This may be expressed by the statement that there is only one kind of temperature, regardless of the variety of scales in which it is expressed. Another modern expression of this idea is that "All diathermal walls are equivalent". This might also be expressed by saying that there is precisely one kind of non-mechanical, non-matter-transferring contact equilibrium between thermodynamic systems.

These ideas may be regarded as helping to clarify the physical meaning of the usual statement of the zeroth law of thermodynamics. It is the opinion of Lieb and Yngvason that the derivation from statistical mechanics of the law of entropy increase is a goal that has so far eluded the deepest thinkers. Thus the idea remains open to consideration that the existence of heat and temperature are needed as coherent primitive concepts for thermodynamics, as expressed, for example, by Maxwell and Planck. On the other hand, Planck in 1926 clarified how the second law can be stated without reference to heat or temperature, by referring to the irreversible and universal nature of friction in natural thermodynamic processes.

First Law of Thermodynamics

The first law of thermodynamics is a version of the law of conservation of energy, adapted for thermodynamic systems. The law of conservation of energy states that the total energy of an isolated system is constant; energy can be transformed from one form to another, but can be neither created nor destroyed. The first law is often formulated:

$$\Delta U = Q - W.$$

It states that the change in the internal energy ΔU of a closed system is equal to the amount of heat Q supplied *to* the system, minus the amount of work W done *by* the system on its surroundings. An equivalent statement is that perpetual motion machines of the first kind are impossible.

Original Statements: The Thermodynamic Approach

The original nineteenth century statements of the first law of thermodynamics appeared in a conceptual framework in which transfer of energy as heat was taken as a primitive notion, not defined or constructed by the theoretical development of the framework, but rather presupposed as prior to it and already accepted. The primitive notion of heat was taken as empirically established, especially through calorimetry regarded as a subject in its own right, prior to thermodynamics. Jointly primitive with this notion of heat were the notions of empirical temperature and thermal equilibrium. This framework also took as primitive the notion of transfer of energy as work. This framework did not presume a concept of energy in general, but regarded it as derived or synthesized from the prior notions of heat and work. By one author, this framework has been called the "thermodynamic" approach.

The first explicit statement of the first law of thermodynamics, by Rudolf Clausius in 1850, referred to cyclic thermodynamic processes.

In all cases in which work is produced by the agency of heat, a quantity of heat is consumed which is proportional to the work done; and conversely, by the expenditure of an equal quantity of work an equal quantity of heat is produced.

Clausius also stated the law in another form, referring to the existence of a function of state of the system, the internal energy, and expressed it in terms of a differential equation for the increments of a thermodynamic process. This equation may be described as follows:

> "In a thermodynamic process involving a closed system, the increment in the internal energy is equal to the difference between the heat accumulated by the system and the work done by it."

Because of its definition in terms of increments, the value of the internal energy of a system is not uniquely defined. It is defined only up to an arbitrary additive constant of integration, which can be adjusted to give arbitrary reference zero levels. This non-uniqueness is in keeping with the abstract mathematical nature of the internal energy. The internal energy is customarily stated relative to a conventionally chosen standard reference state of the system.

The concept of internal energy is considered by Bailyn to be of "enormous interest". Its quantity cannot be immediately measured, but can only be inferred, by differencing actual immediate measurements. Bailyn likens it to the energy states of an atom, that were revealed by Bohr's energy relation $hv = E_{n''} - E_{n'}$. In each case, an unmeasurable quantity (the internal energy, the atomic energy level) is revealed by considering the difference of measured quantities (increments of internal energy, quantities of emitted or absorbed radiative energy).

Conceptual Revision: The Mechanical Approach

In 1907, George H. Bryan wrote about systems between which there is no transfer of matter (closed systems): "Definition. When energy flows from one system or part of a system to another otherwise than by the performance of mechanical work, the energy so transferred is called *heat*." This definition may be regarded as expressing a conceptual revision, as follows. This was systematically expounded in 1909 by Constantin Carathéodory, whose attention had been drawn to it by Max Born. Largely through Born's influence, this revised conceptual approach to the definition of heat came to be preferred by many twentieth-century writers. It might be called the "mechanical approach".

Energy can also be transferred from one thermodynamic system to another in association with transfer of matter. Born points out that in general such energy transfer is not resolvable uniquely into work and heat moieties. In general, when there is transfer of energy associated with matter transfer, work and heat transfers can be distinguished only when they pass through walls physically separate from those for matter transfer.

The "mechanical" approach postulates the law of conservation of energy. It also postulates that energy can be transferred from one thermodynamic system to another adiabatically as work, and that energy can be held as the internal energy of a thermodynamic system. It also postulates that energy can be transferred from one thermodynamic system to another by a path that is non-adiabatic, and is unaccompanied by matter transfer. Initially, it "cleverly" (according to Bailyn) refrains from labelling as 'heat' such non-adiabatic, unaccompanied transfer of energy. It rests on the primitive notion of *walls*, especially adiabatic walls and non-adiabatic walls, defined as follows. Temporarily, only for purpose of this definition, one can prohibit transfer of energy as work across a wall of interest. Then walls of interest fall into two classes, (a) those such that arbitrary systems separated by them remain independently in their own previously established respective states of internal thermodynamic equilibrium; they are defined as adiabatic; and (b) those without such independence; they are defined as non-adiabatic.

This approach derives the notions of transfer of energy as heat, and of temperature, as theoretical developments, not taking them as primitives. It regards calorimetry as a derived theory. It has an early origin in the nineteenth century, for example in the work of Helmholtz, but also in the work of many others.

Conceptually Revised Statement

The revised statement of the first law postulates that a change in the internal energy of a system due to any arbitrary process, that takes the system from a given initial thermodynamic state to a given final equilibrium thermodynamic state, can be determined through the physical existence, for those given states, of a reference process that occurs purely through stages of adiabatic work.

The revised statement is then:

> "For a closed system, in any arbitrary process of interest that takes it from an initial to a final state of internal thermodynamic equilibrium, the change of internal energy is the same as that for a reference adiabatic work process that links those two states. This is so regardless of the path of the process of interest, and regardless of whether it is an adiabatic or a non-adiabatic process. The reference adiabatic work process may be chosen arbitrarily from amongst the class of all such processes."

This statement is much less close to the empirical basis than are the original statements, but is often regarded as conceptually parsimonious in that it rests only on the concepts of adiabatic work and of non-adiabatic processes, not on the concepts of transfer of energy as heat and of empirical temperature that are presupposed by the original statements. Largely through the influence of Max Born, it is often regarded as theoretically preferable because of this conceptual parsimony. Born particularly observes that the revised approach avoids thinking in terms of what he calls the "imported engineering" concept of heat engines.

Basing his thinking on the mechanical approach, Born in 1921, and again in 1949, proposed to revise the definition of heat. In particular, he referred to the work of Constantin Carathéodory, who had in 1909 stated the first law without defining quantity of heat. Born's definition was specifically for transfers of energy without transfer of matter, and it has been widely followed in textbooks. Born observes that a transfer of matter between two systems is accompanied by a transfer of internal energy that cannot be resolved into heat and work components. There can be pathways to other systems, spatially separate from that of the matter transfer, that allow heat and work transfer independent of and simultaneous with the matter transfer. Energy is conserved in such transfers.

Cyclic Processes

The first law of thermodynamics for a closed system was expressed in two ways by Clausius. One way referred to cyclic processes and the inputs and outputs of the system, but did not refer to increments in the internal state of the system. The other way referred to an incremental change in the internal state of the system, and did not expect the process to be cyclic.

A cyclic process is one that can be repeated indefinitely often, returning the system to its initial state. Of particular interest for single cycle of a cyclic process are the net work done, and the net heat taken in (or 'consumed', in Clausius' statement), by the system.

In a cyclic process in which the system does net work on its surroundings, it is observed to be physically necessary not only that heat be taken into the system, but also, importantly, that some heat leave the system. The difference is the heat converted by the cycle into work. In each repetition of a cyclic process, the net work done by the system, measured in mechanical units, is proportional to the heat consumed, measured in calorimetric units.

The constant of proportionality is universal and independent of the system and in 1845 and 1847 was measured by James Joule, who described it as the *mechanical equivalent of heat.*

Sign Conventions

In a non-cyclic process, the change in the internal energy of a system is equal to net energy added as heat to the system minus the net work done by the system, both being measured in mechanical units. Taking ΔU as a change in internal energy, one writes:

$$\Delta U = Q - W$$

where Q denotes the net quantity of heat supplied to the system by its surroundings and W denotes the net work done by the system. This sign convention is implicit in Clausius' statement of the law given above. It originated with the study of heat engines that produce useful work by consumption of heat.

Often nowadays, however, writers use the IUPAC convention by which the first law is formulated with work done on the system by its surroundings having a positive sign. With this now often used sign convention for work, the first law for a closed system may be written:

$$\Delta U = Q + W$$

This convention follows physicists such as Max Planck, and considers all net energy transfers to the system as positive and all net energy transfers from the system as negative, irrespective of any use for the system as an engine or other device.

When a system expands in a fictive quasistatic process, the work done by the system on the environment is the product, $P\,\mathrm{d}V$, of pressure, P, and volume change, $\mathrm{d}V$, whereas the work done *on* the system is $-P\,\mathrm{d}V$. Using either sign convention for work, the change in internal energy of the system is:

$$\mathrm{d}U = \delta Q - P\mathrm{d}V$$

where δQ denotes the infinitesimal amount of heat supplied to the system from its surroundings and δ denotes an inexact differential.

Work and heat are expressions of actual physical processes of supply or removal of energy, while the internal energy U is a mathematical abstraction that keeps account of the exchanges of energy that befall the system. Thus the term heat for Q means "that amount of energy added or removed by conduction of heat or by thermal radiation", rather than referring to a form of energy within the system. Likewise, the term work energy for W means "that amount of energy gained or lost as the result of work". Internal energy is a property of the system whereas work done and heat supplied are not. A significant result of this distinction is that a given internal energy change ΔU can be achieved by, in principle, many combinations of heat and work.

Various Statements of the Law for Closed Systems

The law is of great importance and generality and is consequently thought of from several points of view. Most careful textbook statements of the law express it for closed systems. It is stated in several ways, sometimes even by the same author.

For the thermodynamics of closed systems, the distinction between transfers of energy as work and as heat is.

There are two main ways of stating a law of thermodynamics, physically or mathematically. They should be logically coherent and consistent with one another.

An example of a physical statement is that of Planck:

> "It is in no way possible, either by mechanical, thermal, chemical, or other devices, to obtain perpetual motion, i.e. it is impossible to construct an engine which will work in a cycle and produce continuous work, or kinetic energy, from nothing."

This physical statement is restricted neither to closed systems nor to systems with states that are strictly defined only for thermodynamic equilibrium; it has meaning also for open systems and for systems with states that are not in thermodynamic equilibrium.

An example of a mathematical statement is that of Crawford:

> "For a given system we let ΔE^{kin} = large-scale mechanical energy, ΔE^{pot} = large-scale potential energy, and ΔE^{tot} = total energy. The first two quantities are specifiable in terms of appropriate mechanical variables, and by definition
>
> $$E^{\mathrm{tot}} = E^{\mathrm{kin}} + E^{\mathrm{pot}} + U.$$
>
> For any finite process, whether reversible or irreversible,
>
> $$\Delta E^{\mathrm{tot}} = \Delta E^{\mathrm{kin}} + \Delta E^{\mathrm{pot}} + \Delta U.$$
>
> The first law in a form that involves the principle of conservation of energy more generally is
>
> $$\Delta E^{\mathrm{tot}} = Q + W.$$
>
> Here Q and W are heat and work added, with no restrictions as to whether the process is reversible, quasistatic, or irreversible."

This statement by Crawford, for W, uses the sign convention of IUPAC, not that of Clausius. Though it does not explicitly say so, this statement refers to closed systems, and to internal energy U defined for bodies in states of thermodynamic equilibrium, which possess well-defined temperatures.

The history of statements of the law for closed systems has two main periods, before and after the work of Bryan, of Carathéodory, and the approval of Carathéodory's work given by Born. The earlier traditional versions of the law for closed systems are nowadays often considered to be out of date.

Carathéodory's celebrated presentation of equilibrium thermodynamics refers to closed systems, which are allowed to contain several phases connected by internal walls of various kinds of impermeability and permeability (explicitly including walls that are permeable only to heat). Carathéodory's 1909 version of the first law of thermodynamics was stated in an axiom which refrained from defining or mentioning temperature or quantity of heat transferred. That axiom stated that the internal energy of a phase in equilibrium is a function of state, that the sum of the internal energies of the phases is the total internal energy of the system, and that the value of the total internal energy of the system is changed by the amount of work done adiabatically on it, considering work as a form of energy. This version is nowadays widely accepted as authoritative, but is stated in slightly varied ways by different authors.

Such statements of the first law for closed systems assert the existence of internal energy as a function of state defined in terms of adiabatic work. Thus heat is not defined calorimetrically or as due to temperature difference. It is defined as a residual difference between change of internal energy and work done on the system, when that work does not account for the whole of the change of internal energy and the system is not adiabatically isolated.

The 1909 Carathéodory statement of the law in axiomatic form does not mention heat or temperature, but the equilibrium states to which it refers are explicitly defined by variable sets that necessarily include "non-deformation variables", such as pressures, which, within reasonable restrictions, can be rightly interpreted as empirical temperatures, and the walls connecting the phases of the system are explicitly defined as possibly impermeable to heat or permeable only to heat.

According to Münster, "A somewhat unsatisfactory aspect of Carathéodory's theory is that a consequence of the Second Law must be considered at this point (in the statement of the first law), i.e. that it is not always possible to reach any state 2 from any other state 1 by means of an adiabatic process." Münster instances that no adiabatic process can reduce the internal energy of a system at constant volume. Carathéodory's paper asserts that its statement of the first law corresponds exactly to Joule's experimental arrangement, regarded as an instance of adiabatic work. It does not point out that Joule's experimental arrangement performed essentially irreversible work, through friction of paddles in a liquid, or passage of electric current through a resistance inside the system, driven by motion of a coil and inductive heating, or by an external current source, which can access the system only by the passage of electrons, and so is not strictly adiabatic, because electrons are a form of matter, which cannot penetrate adiabatic walls. The paper goes on to base its main argument on the possibility of quasi-static adiabatic work, which is essentially reversible. The paper asserts that it will avoid reference to Carnot cycles, and then proceeds to base its argument on cycles of forward and backward quasi-static adiabatic stages, with isothermal stages of zero magnitude.

Sometimes the concept of internal energy is not made explicit in the statement. Sometimes the existence of the internal energy is made explicit but work is not explicitly mentioned in the statement of the first postulate of thermodynamics. Heat supplied is then defined as the residual change in internal energy after work has been taken into account, in a non-adiabatic process.

A respected modern author states the first law of thermodynamics as "Heat is a form of energy", which explicitly mentions neither internal energy nor adiabatic work. Heat is defined as energy transferred by thermal contact with a reservoir, which has a temperature, and is generally so large that addition and removal of heat do not alter its temperature. A current student text on chemistry defines heat thus: "*heat* is the exchange of thermal energy between a system and its surroundings caused by a temperature difference." The author then explains how heat is defined or measured by calorimetry, in terms of heat capacity, specific heat capacity, molar heat capacity, and temperature.

A respected text disregards the Carathéodory's exclusion of mention of heat from the statement of the first law for closed systems, and admits heat calorimetrically defined along with work and internal energy. Another respected text defines heat exchange as determined by temperature difference, but also mentions that the Born version is "completely rigorous". These versions follow the traditional approach that is now considered out of date, exemplified by that of Planck.

Evidence for the First Law of Thermodynamics for Closed Systems

The first law of thermodynamics for closed systems was originally induced from empirically observed evidence, including calorimetric evidence. It is nowadays, however, taken to provide the definition of heat via the law of conservation of energy and the definition of work in terms of changes in the external parameters of a system. The original discovery of the law was gradual over a period of perhaps half a century or more, and some early studies were in terms of cyclic processes.

The following is an account in terms of changes of state of a closed system through compound processes that are not necessarily cyclic. This account first considers processes for which the first law is easily verified because of their simplicity, namely adiabatic processes (in which there is no transfer as heat) and adynamic processes (in which there is no transfer as work).

Adiabatic Processes

In an adiabatic process, there is transfer of energy as work but not as heat. For all adiabatic process that takes a system from a given initial state to a given final state, irrespective of how the work is done, the respective eventual total quantities of energy transferred as work are one and the same, determined just by the given initial and final states. The work done on the system is defined and measured by changes in mechanical or quasi-mechanical variables external to the system. Physically, adiabatic transfer of energy as work requires the existence of adiabatic enclosures.

For instance, in Joule's experiment, the initial system is a tank of water with a paddle wheel inside. If we isolate the tank thermally, and move the paddle wheel with a pulley and a weight, we can relate the increase in temperature with the distance descended by the mass. Next, the system is returned to its initial state, isolated again, and the same amount of work is done on the tank using different devices (an electric motor, a chemical battery, a spring). In every case, the amount of work can be measured independently. The return to the initial state is not conducted by doing adiabatic work on the system. The evidence shows that the final state of the water (in particular, its temperature and volume) is the same in every case. It is irrelevant if the work is electrical, mechanical, chemical or if done suddenly or slowly, as long as it is performed in an adiabatic way, that is to say, without heat transfer into or out of the system.

Evidence of this kind shows that to increase the temperature of the water in the tank, the qualitative kind of adiabatically performed work does not matter. No qualitative kind of adiabatic work has ever been observed to decrease the temperature of the water in the tank.

A change from one state to another, for example an increase of both temperature and volume, may be conducted in several stages, for example by externally supplied electrical work on a resistor in the body, and adiabatic expansion allowing the body to do work on the surroundings. It needs to be shown that the time order of the stages, and their relative magnitudes, does not affect the amount of adiabatic work that needs to be done for the change of state. According to one respected scholar: "Unfortunately, it does not seem that experiments of this kind have ever been carried out carefully. We must therefore admit that the statement which we have enunciated here, and which is equivalent to the first law of thermodynamics, is not well founded on direct experimental evidence." Another expression of this view is "no systematic precise experiments to verify this generalization directly have ever been attempted."

This kind of evidence, of independence of sequence of stages, combined with the above-mentioned evidence, of independence of qualitative kind of work, would show the existence of an important state variable that corresponds with adiabatic work, but not that such a state variable represented a conserved quantity. For the latter, another step of evidence is needed, which may be related to the concept of reversibility, as mentioned below.

That important state variable was first recognized and denoted U by Clausius in 1850, but he did not then name it, and he defined it in terms not only of work but also of heat transfer in the same process. It was also independently recognized in 1850 by Rankine, who also denoted it U ; and in 1851 by Kelvin who then called it "mechanical energy", and later "intrinsic energy". In 1865, after some hestitation, Clausius began calling his state function U "energy". In 1882 it was named as the *internal energy* by Helmholtz. If only adiabatic processes were of interest, and heat could be ignored, the concept of internal energy would hardly arise or be needed. The relevant physics would be largely covered by the concept of potential energy, as was intended in the 1847 paper of Helmholtz on the principle of conservation of energy, though that did not deal with forces that cannot be described by a potential, and thus did not fully justify the principle. Moreover, that paper was critical of the early work of Joule that had by then been performed. A great merit of the internal energy concept is that it frees thermodynamics from a restriction to cyclic processes, and allows a treatment in terms of thermodynamic states.

In an adiabatic process, adiabatic work takes the system either from a reference state O with internal energy $U(O)$ to an arbitrary one A with internal energy $U(A)$, or from the state A to the state O:

$$U(A) = U(O) - W_{O \to A}^{\text{adiabatic}} \text{ or } U(O) = U(A) - W_{A \to O}^{\text{adiabatic}}.$$

Except under the special, and strictly speaking, fictional, condition of reversibility, only one of the processes adiabatic, $O \to A$ or adiabatic, $A \to O$ is empirically feasible by a simple application of externally supplied work.

The fact of such irreversibility may be dealt with in two main ways, according to different points of view:

- Since the work of Bryan, the most accepted way to deal with it nowadays, followed by Carathéodory, is to rely on the previously established concept of quasi-static processes, as follows. Actual physical processes of transfer of energy as work are always at least to some degree irreversible. The irreversibility is often due to mechanisms known as dissipative, that transform bulk kinetic energy into internal energy. Examples are friction and viscosity. If the process is performed more slowly, the frictional or viscous dissipation is less. In the limit of infinitely slow performance, the dissipation tends to zero and then the limiting process, though fictional rather than actual, is notionally reversible, and is called quasi-static. Throughout the course of the fictional limiting quasi-static process, the internal intensive variables of the system are equal to the external intensive variables, those that describe the reactive forces exerted by the surroundings. This can be taken to justify the formula:

$$W_{A\to O}^{\text{adiabatic, quasi-static}} = -W_{O\to A}^{\text{adiabatic, quasi-static}}.$$

- Another way to deal with it is to allow that experiments with processes of heat transfer to or from the system may be used to justify the formula above. Moreover, it deals to some extent with the problem of lack of direct experimental evidence that the time order of stages of a process does not matter in the determination of internal energy. This way does not provide theoretical purity in terms of adiabatic work processes, but is empirically feasible, and is in accord with experiments actually done, such as the Joule experiments mentioned just above, and with older traditions.

The formula above allows that to go by processes of quasi-static adiabatic work from the state A to the state B we can take a path that goes through the reference state O, since the quasi-static adiabatic work is independent of the path:

$$-W_{A\to B}^{\text{adiabatic,quasi-static}} = -W_{A\to O}^{\text{adiabatic,quasi-static}} - W_{O\to B}^{\text{adiabatic,quasi-static}} = W_{O\to A}^{\text{adiabatic,quasi-static}} - W_{O\to B}^{\text{adiabatic,quasi-static}}$$

$$= -U(A) + U(B) = \Delta U$$

This kind of empirical evidence, coupled with theory of this kind, largely justifies the following statement:

> "For all adiabatic processes between two specified states of a closed system of any nature, the net work done is the same regardless the details of the process, and determines a state function called internal energy, U."

Adynamic Processes

A complementary observable aspect of the first law is about heat transfer. Adynamic transfer of energy as heat can be measured empirically by changes in the surroundings of the system of interest by calorimetry. This again requires the existence of adiabatic enclosure of the entire process, system and surroundings, though the separating wall between the surroundings and the system is thermally conductive or radiatively permeable, not adiabatic. A calorimeter can rely on measurement of sensible heat, which requires the existence of thermometers and measurement of temperature change in bodies of known sensible heat capacity under specified conditions; or it can rely on the measurement of latent heat, through measurement of masses of material that change phase, at temperatures fixed by the occurrence of phase changes under specified conditions in bodies of known latent heat of phase change. The calorimeter can be calibrated by adiabatically doing externally determined work on it. The most accurate method is by passing an electric current from outside through a resistance inside the calorimeter. The calibration allows comparison of calorimetric measurement of quantity of heat transferred with quantity of energy transferred as work. According to one textbook, "The most common device for measuring ΔU is an adiabatic bomb calorimeter". According to another textbook, "Calorimetry is widely used in present day laboratories". According to one opinion, "Most thermodynamic data come from calorimetry". According to another opinion, "The most common method of measuring "heat" is with a calorimeter."

When the system evolves with transfer of energy as heat, without energy being transferred as work, in an adynamic process, the heat transferred to the system is equal to the increase in its internal energy:

$$Q_{A\to B}^{\text{adynamic}} = \Delta U.$$

General Case for Reversible Processes

Heat transfer is practically reversible when it is driven by practically negligibly small temperature gradients. Work transfer is practically reversible when it occurs so slowly that there are no frictional effects within the system; frictional effects outside the system should also be zero if the process is to be globally reversible. For a particular reversible process in general, the work done reversibly on the system, $W_{A\to B}^{\text{path}\,P_0,\text{reversible}}$, and the heat transferred reversibly to the system, $Q_{A\to B}^{\text{path}\,P_0,\text{reversible}}$ are not required to occur respectively adiabatically or adynamically, but they must belong to the same particular process defined by its particular reversible path, P_0, through the space of thermodynamic states. Then the work and heat transfers can occur and be calculated simultaneously.

Putting the two complementary aspects together, the first law for a particular reversible process can be written

$$-W_{A\to B}^{\text{path}\,P_0,\text{reversible}} + Q_{A\to B}^{\text{path}\,P_0,\text{reversible}} = \Delta U.$$

This combined statement is the expression the first law of thermodynamics for reversible processes for closed systems.

In particular, if no work is done on a thermally isolated closed system we have

$$\Delta U = 0.$$

This is one aspect of the law of conservation of energy and can be stated:

> "The internal energy of an isolated system remains constant."

General Case for Irreversible Processes

If, in a process of change of state of a closed system, the energy transfer is not under a practically zero temperature gradient and practically frictionless, then the process is irreversible. Then the heat and work transfers may be difficult to calculate, and irreversible thermodynamics is called for. Nevertheless, the first law still holds and provides a check on the measurements and calculations of the work done irreversibly on the system, $W_{A\to B}^{\text{path}\,P_1,\text{irreversible}}$, and the heat transferred irreversibly to the system, $Q_{A\to B}^{\text{path } P_1,\text{irreversible}}$, which belong to the same particular process defined by its particular irreversible path, P_1, through the space of thermodynamic states.

$$-W_{A\to B}^{\text{path } P_1,\text{irreversible}} + Q_{A\to B}^{\text{path } P_1,\text{irreversible}} = \Delta U.$$

This means that the internal energy U is a function of state and that the internal energy change ΔU between two states is a function only of the two states.

Weight of Evidence for the Law

The first law of thermodynamics is so general that its predictions cannot all be directly tested. In many properly conducted experiments it has been precisely supported, and never violated. Indeed, within its scope of applicability, the law is so reliably established, that, nowadays, rather than experiment being considered as testing the accuracy of the law, it is more practical and realistic to think of the law as testing the accuracy of experiment. An experimental result that seems to violate the law may be assumed to be inaccurate or wrongly conceived, for example due to failure to account for an important physical factor. Thus, some may regard it as a principle more abstract than a law.

State Functional Formulation for Infinitesimal Processes

When the heat and work transfers in the equations above are infinitesimal in magnitude, they are often denoted by δ, rather than exact differentials denoted by d, as a reminder that heat and work do not describe the *state* of any system. The integral of an inexact differential depends upon the particular path taken through the space of thermodynamic parameters while the integral of an exact differential depends only upon the initial and final states. If the initial and final states are the same, then the integral of an inexact differential may or may not be zero, but the integral of an exact differential is always zero. The path taken by a thermodynamic system through a chemical or physical change is known as a thermodynamic process.

The first law for a closed homogeneous system may be stated in terms that include concepts that are established in the second law. The internal energy U may then be expressed as a function of the system's defining state variables S, entropy, and V, volume: $U = U(S, V)$. In these terms, T, the system's temperature, and P, its pressure, are partial derivatives of U with respect to S and V. These variables are important throughout thermodynamics, though not necessary for the statement of the first law. Rigorously, they are defined only when the system is in its own state of internal thermodynamic equilibrium. For some purposes, the concepts provide good approximations for scenarios sufficiently near to the system's internal thermodynamic equilibrium.

The first law requires that:

$$dU = \delta Q - \delta W \qquad \text{(closed system, general process, quasi-static or irreversible).}$$

Then, for the fictive case of a reversible process, dU can be written in terms of exact differentials. One may imagine reversible changes, such that there is at each instant negligible departure from thermodynamic equilibrium within the system. This excludes isochoric work. Then, mechanical work is given by $\delta W = -P\,dV$ and the quantity of heat added can be expressed as $\delta Q = T\,dS$. For these conditions:

$$dU = TdS - PdV \qquad \text{(closed system, reversible process).}$$

While this has been shown here for reversible changes, it is valid in general, as U can be considered as a thermodynamic state function of the defining state variables S and V:

$$dU = TdS - PdV \text{ (closed system, general process, quasi-static or irreversible).}$$

Equation above is known as the fundamental thermodynamic relation for a closed system in the energy representation, for which the defining state variables are S and V, with respect to which T and P are partial derivatives of U. It is only in the fictive reversible case, when isochoric work is excluded, that the work done and heat transferred are given by $-P\,dV$ and $T\,dS$.

In the case of a closed system in which the particles of the system are of different types and, because chemical reactions may occur, their respective numbers are not necessarily constant, the fundamental thermodynamic relation for dU becomes:

$$dU = TdS - PdV + \sum_i \mu_i dN_i.$$

where dN_i is the (small) increase in number of type-i particles in the reaction, and μ_i is known as the chemical potential of the type-i particles in the system. If dN_i is expressed in mol then μ_i is expressed in J/mol. If the system has more external mechanical variables than just the volume that can change, the fundamental thermodynamic relation further generalizes to:

$$dU = TdS - \sum_i X_i dx_i + \sum_j \mu_j dN_j.$$

Here the X_i are the generalized forces corresponding to the external variables x_i. The parameters X_i are independent of the size of the system and are called intensive parameters and the x_i are proportional to the size and called extensive parameters.

For an open system, there can be transfers of particles as well as energy into or out of the system during a process. For this case, the first law of thermodynamics still holds, in the form that the internal energy is a function of state and the change of internal energy in a process is a function only of its initial and final states.

A useful idea from mechanics is that the energy gained by a particle is equal to the force applied to the particle multiplied by the displacement of the particle while that force is applied. Now consider the first law without the heating term: dU = $-P$dV. The pressure P can be viewed as a force (and in fact has units of force per unit area) while dV is the displacement (with units of distance times area). We may say, with respect to this work term, that a pressure difference forces a transfer of volume, and that the product of the two (work) is the amount of energy transferred out of the system as a result of the process. If one were to make this term negative then this would be the work done on the system.

It is useful to view the TdS term in the same light: here the temperature is known as a "generalized" force (rather than an actual mechanical force) and the entropy is a generalized displacement.

Similarly, a difference in chemical potential between groups of particles in the system drives a chemical reaction that changes the numbers of particles, and the corresponding product is the amount of chemical potential energy transformed in process. For example, consider a system consisting of two phases: liquid water and water vapor. There is a generalized "force" of evaporation that drives water molecules out of the liquid. There is a generalized "force" of condensation that drives vapor molecules out of the vapor. Only when these two "forces" (or chemical potentials) are equal is there equilibrium, and the net rate of transfer zero.

The two thermodynamic parameters that form a generalized force-displacement pair are called "conjugate variables". The two most familiar pairs are, of course, pressure-volume, and temperature-entropy.

Spatially Inhomogeneous Systems

Classical thermodynamics is initially focused on closed homogeneous systems, which might be regarded as 'zero-dimensional' in the sense that they have no spatial variation. But it is desired to study also systems with distinct internal motion and spatial inhomogeneity. For such systems, the principle of conservation of energy is expressed in terms not only of internal energy as defined for homogeneous systems, but also in terms of kinetic energy and potential energies of parts of the inhomogeneous system with respect to each other and with respect to long-range external forces. How the total energy of a system is allocated between these three more specific kinds of energy varies according to the purposes of different writers; this is because these components of energy are to some extent mathematical artefacts rather than actually measured physical quantities. For any closed homogeneous component of an inhomogeneous closed system, if E denotes the total energy of that component system, one may write:

$$E = E^{\mathrm{kin}} + E^{\mathrm{pot}} + U$$

where E^{kin} and E^{pot} denote respectively the total kinetic energy and the total potential energy of the component closed homogeneous system, and U denotes its internal energy.

Potential energy can be exchanged with the surroundings of the system when the surroundings impose a force field, such as gravitational or electromagnetic, on the system.

A compound system consisting of two interacting closed homogeneous component subsystems has a potential energy of interaction E_{12}^{pot} between the subsystems. Thus, in an obvious notation, one may write:

$$E = E_1^{\mathrm{kin}} + E_1^{\mathrm{pot}} + U_1 + E_2^{\mathrm{kin}} + E_2^{\mathrm{pot}} + U_2 + E_{12}^{\mathrm{pot}}$$

The quantity E_{12}^{pot} in general lacks an assignment to either subsystem in a way that is not arbitrary, and this stands in the way of a general non-arbitrary definition of transfer of energy as work. On occasions, authors make their various respective arbitrary assignments.

The distinction between internal and kinetic energy is hard to make in the presence of turbulent motion within the system, as friction gradually dissipates macroscopic kinetic energy of localised bulk flow into molecular random motion of molecules that is classified as internal energy. The rate of dissipation by friction of kinetic energy of localised bulk flow into internal energy, whether in turbulent or in streamlined flow, is an important quantity in non-equilibrium thermodynamics. This is a serious difficulty for attempts to define entropy for time-varying spatially inhomogeneous systems.

First Law of Thermodynamics for Open Systems

For the first law of thermodynamics, there is no trivial passage of physical conception from the closed system view to an open system view. For closed systems, the concepts of an adiabatic enclosure and of an adiabatic wall are fundamental. Matter and internal energy cannot permeate or penetrate such a wall. For an open system, there is a wall that allows penetration by matter. In general, matter in diffusive motion carries with it some internal energy, and some microscopic potential energy changes accompany the motion. An open system is not adiabatically enclosed.

There are some cases in which a process for an open system can, for particular purposes, be considered as if it were for a closed system. In an open system, by definition hypothetically or potentially, matter can pass between the system and its surroundings. But when, in a particular case, the process of interest involves only hypothetical or potential but no actual passage of matter, the process can be considered as if it were for a closed system.

Internal Energy for an Open System

Since the revised and more rigorous definition of the internal energy of a closed system rests upon the possibility of processes by which adiabatic work takes the system from one state to another, this leaves a problem for the definition of internal energy for an open system, for which adiabatic work is not in general possible. According to Max Born, the transfer of matter and energy across an open connection "cannot be reduced to mechanics". In contrast to the case of closed systems, for open systems, in the presence of diffusion, there is no unconstrained and unconditional physical distinction between convective transfer of internal energy by bulk flow of matter, the transfer of internal energy without transfer of matter (usually called heat conduction and work transfer), and change of various potential energies. The older traditional way and the conceptually revised (Carathéodory) way agree that there is no physically unique definition of heat and work transfer processes between open systems.

In particular, between two otherwise isolated open systems an adiabatic wall is by definition impossible. This problem is solved by recourse to the principle of conservation of energy. This principle allows a composite isolated system to be derived from two other component non-interacting isolated systems, in such a way that the total energy of the composite isolated system is equal to the sum of the total energies of the two component isolated systems. Two previously isolated systems can be subjected to the thermodynamic operation of placement between them of a wall permeable to matter and energy, followed by a time for establishment of a new thermodynamic state of internal equilibrium in the new single unpartitioned system. The internal energies of the initial two systems and of the final new system, considered respectively as closed systems as above, can be measured. Then the law of conservation of energy requires that:

$$\Delta U_s + \Delta U_o = 0,$$

where ΔU_s and ΔU_0 denote the changes in internal energy of the system and of its surroundings respectively. This is a statement of the first law of thermodynamics for a transfer between two otherwise isolated open systems, that fits well with the conceptually revised and rigorous statement of the law stated above.

For the thermodynamic operation of adding two systems with internal energies U_1 and U_2, to produce a new system with internal energy U, one may write $U = U_1 + U_2$; the reference states for U, U_1 and U_2 should be specified accordingly, maintaining also that the internal energy of a system be proportional to its mass, so that the internal energies are extensive variables.

There is a sense in which this kind of additivity expresses a fundamental postulate that goes beyond the simplest ideas of classical closed system thermodynamics; the extensivity of some variables is not obvious, and needs explicit expression; indeed one author goes so far as to say that it could be recognized as a fourth law of thermodynamics, though this is not repeated by other authors.

Also of course:

$$\Delta N_s + \Delta N_o = 0,$$

where ΔN_s and ΔN_o denote the changes in mole number of a component substance of the system and of its surroundings respectively. This is a statement of the law of conservation of mass.

Transfer of Matter between an Open System and its Surroundings

A system connected to its surroundings only through contact by a single permeable wall, but otherwise isolated, is an open system. If it is initially in a state of contact equilibrium with a surrounding subsystem, a thermodynamic process of transfer of matter can be made to occur between them if the surrounding subsystem is subjected to some thermodynamic operation, for example, removal of a partition between it and some further surrounding subsystem. The removal of the partition in the surroundings initiates a process of exchange between the system and its contiguous surrounding subsystem.

An example is evaporation. One may consider an open system consisting of a collection of liquid, enclosed except where it is allowed to evaporate into or to receive condensate from its vapor above it, which may be considered as its contiguous surrounding subsystem, and subject to control of its volume and temperature.

A thermodynamic process might be initiated by a thermodynamic operation in the surroundings, that mechanically increases in the controlled volume of the vapor. Some mechanical work will be done within the surroundings by the vapor, but also some of the parent liquid will evaporate and enter the vapor collection which is the contiguous surrounding subsystem. Some internal energy will accompany the vapor that leaves the system, but it will not make sense to try to uniquely identify part of that internal energy as heat and part of it as work. Consequently, the energy transfer that accompanies the transfer of matter between the system and its surrounding subsystem cannot be uniquely split into heat and work transfers to or from the open system. The component of total energy transfer that accompanies the transfer of vapor into the surrounding subsystem is customarily called 'latent heat of evaporation', but this use of the word heat is a quirk of customary historical language, not in strict compliance with the thermodynamic definition of transfer of energy as heat. In this example, kinetic energy of bulk flow and potential energy with respect to long-range external forces such as gravity are both considered to be zero. The first law of thermodynamics refers to the change of internal energy of the open system, between its initial and final states of internal equilibrium.

Open System with Multiple Contacts

An open system can be in contact equilibrium with several other systems at once.

This includes cases in which there is contact equilibrium between the system, and several subsystems in its surroundings, including separate connections with subsystems through walls that are permeable to the transfer of matter and internal energy as heat and allowing friction of passage of the transferred matter, but immovable, and separate connections through adiabatic walls with others, and separate connections through diathermic walls impermeable to matter with yet others. Because there are physically separate connections that are permeable to energy but impermeable to matter, between the system and its surroundings, energy transfers between them can occur with definite heat and work characters. Conceptually essential here is that the internal energy transferred with the transfer of matter is measured by a variable that is mathematically independent of the variables that measure heat and work.

With such independence of variables, the total increase of internal energy in the process is then determined as the sum of the internal energy transferred from the surroundings with the transfer of matter through the walls that are permeable to it, and of the internal energy transferred to the system as heat through the diathermic walls, and of the energy transferred to the system as work through the adiabatic walls, including the energy transferred to the system by long-range forces. These simultaneously transferred quantities of energy are defined by events in the surroundings of the system. Because the internal energy transferred with matter is not in general uniquely resolvable into heat and work components, the total energy transfer cannot in general be uniquely resolved into heat and work components. Under these conditions, the following formula can describe the process in terms of externally defined thermodynamic variables, as a statement of the first law of thermodynamics:

$$\Delta U_0 = Q - W - \sum_{i=1}^{m} \Delta U_i \quad \text{(suitably defined surrounding subsystems, general process, quasi-static or irreversible)},$$

where ΔU_0 denotes the change of internal energy of the system, and ΔU_i denotes the change of internal energy of the ith of the m surrounding subsystems that are in open contact with the system, due to transfer between the system and that ith surrounding subsystem, and Q denotes the internal energy transferred as heat from the heat reservoir of the surroundings to the system, and W denotes the energy transferred from the system to the surrounding subsystems that are in adiabatic connection with it. The case of a wall that is permeable to matter and can move so as to allow transfer of energy as work is not considered here.

Combination of First and Second Laws

If the system is described by the energetic fundamental equation, $U_0 = U_0(S, V, N_j)$, and if the process can be described in the quasi-static formalism, in terms of the internal state variables of the system, then the process can also be described by a combination of the first and second laws of thermodynamics, by the formula:

$$dU_0 = T\,dS - P\,dV + \sum_{j=1}^{n} \mu_j\,dN_j$$

where there are n chemical constituents of the system and permeably connected surrounding subsystems, and where T, S, P, V, N_j, and μ_j, are defined as above.

For a general natural process, there is no immediate term-wise correspondence between equations $\Delta U_0 = Q - W - \sum_{i=1}^{m} \Delta U_i$ and $dU_0 = TdS - PdV + \sum_{j=1}^{n} \mu_j dN_j$, because they describe the process in different conceptual frames.

Nevertheless, a conditional correspondence exists. There are three relevant kinds of wall here: purely diathermal, adiabatic, and permeable to matter. If two of those kinds of wall are sealed off, leaving only one that permits transfers of energy, as work, as heat, or with matter, then the remaining permitted terms correspond precisely. If two of the kinds of wall are left unsealed, then energy transfer can be shared between them, so that the two remaining permitted terms do not correspond precisely.

For the special fictive case of quasi-static transfers, there is a simple correspondence. For this, it is supposed that the system has multiple areas of contact with its surroundings. There are pistons that allow adiabatic work, purely diathermal walls, and open connections with surrounding subsystems of completely controllable chemical potential (or equivalent controls for charged species). Then, for a suitable fictive quasi-static transfer, one can write:

$$\delta Q = TdS \text{ and } \delta W = PdV \quad \text{(suitably defined surrounding subsystems, quasi-static transfers of energy).}$$

For fictive quasi-static transfers for which the chemical potentials in the connected surrounding subsystems are suitably controlled, these can be put into equation $dU_0 = TdS - PdV + \sum_{j=1}^{n} \mu_j dN_j$ to yield:

$$dU_0 = \delta Q - \delta W + \sum_{j=1}^{n} \mu_j dN_j \quad \text{(suitably defined surrounding subsystems, quasi-static transfers).}$$

The reference does not actually write in above equation, but what it does write is fully compatible with it. Another helpful account is given by Tschoegl.

There are several other accounts of this, in apparent mutual conflict.

Non-equilibrium Transfers

The transfer of energy between an open system and a single contiguous subsystem of its surroundings is considered also in non-equilibrium thermodynamics. The problem of definition arises also in this case. It may be allowed that the wall between the system and the subsystem is not only permeable to matter and to internal energy, but also may be movable so as to allow work to be done when the two systems have different pressures. In this case, the transfer of energy as heat is not defined.

Methods for study of non-equilibrium processes mostly deal with spatially continuous flow systems. In this case, the open connection between system and surroundings is usually taken to fully surround the system, so that there are no separate connections impermeable to matter but permeable to heat. Except for the special case mentioned above when there is no actual transfer of matter, which can be treated as if for a closed system, in strictly defined thermodynamic terms, it follows that transfer of energy as heat is not defined. In this sense, there is no such thing as 'heat flow' for a continuous-flow open system. Properly, for closed systems, one speaks of transfer of internal energy as heat, but in general, for open systems, one can speak safely only of transfer of internal energy. A factor here is that there are often cross-effects between distinct transfers, for example that transfer of one substance may cause transfer of another even when the latter has zero chemical potential gradient.

Usually transfer between a system and its surroundings applies to transfer of a state variable, and obeys a balance law, that the amount lost by the donor system is equal to the amount gained by the receptor system. Heat is not a state variable. For his 1947 definition of "heat transfer" for discrete open systems, the author Prigogine carefully explains at some length that his definition of it does not obey a balance law. He describes this as paradoxical.

The situation is clarified by Gyarmati, who shows that his definition of "heat transfer", for continuous-flow systems, really refers not specifically to heat, but rather to transfer of internal energy, as follows. He considers a conceptual small cell in a situation of continuous-flow as a system defined in the so-called Lagrangian way, moving with the local center of mass. The flow of matter across the boundary is zero when considered as a flow of total mass. Nevertheless, if the material constitution is of several chemically distinct components that can diffuse with respect to one another, the system is considered to be open, the diffusive flows of the components being defined with respect to the center of mass of the system, and balancing one another as to mass transfer. Still there can be a distinction between bulk flow of internal energy and diffusive flow of internal energy in this case, because the internal energy density does not have to be constant per unit mass of material, and allowing for non-conservation of internal energy because of local conversion of kinetic energy of bulk flow to internal energy by viscosity.

Gyarmati shows that his definition of "the heat flow vector" is strictly speaking a definition of flow of internal energy, not specifically of heat, and so it turns out that his use here of the word heat is contrary to the strict thermodynamic definition of heat, though it is more or less compatible with historical custom, that often enough did not clearly distinguish between heat and internal energy; he writes "that this relation must be considered to be the exact definition of the concept of heat flow, fairly loosely used in experimental physics and heat technics." Apparently in a different frame of thinking from that of the above-mentioned paradoxical usage in the earlier sections of the historic 1947 work by Prigogine, about discrete systems, this usage of Gyarmati is consistent with the later sections of the same 1947 work by Prigogine, about continuous-flow systems, which use the term "heat flux" in just this way. This usage is also followed by Glansdorff and Prigogine in their 1971 text about continuous-flow systems. They write: "Again the flow of internal energy may be split into a convection flow $\rho u v$ and a conduction flow. This conduction flow is by definition the heat flow W. Therefore: $j[U] = \rho u v + W$ where u denotes the internal energy per unit mass. This usage is followed also by other writers on non-equilibrium thermodynamics such as Lebon, Jou, and Casas-Vásquez, and de Groot and Mazur. This usage is described by Bailyn as stating the non-convective flow of internal energy, and is listed as his definition number 1, according to the first law of thermodynamics. This usage is also followed by workers in the kinetic theory of gases. This is not the *ad hoc* definition of "reduced heat flux" of Haase.

In the case of a flowing system of only one chemical constituent, in the Lagrangian representation, there is no distinction between bulk flow and diffusion of matter. Moreover, the flow of matter is zero into or out of the cell that moves with the local center of mass. In effect, in this description, one is dealing with a system effectively closed to the transfer of matter. But still one can validly talk of a distinction between bulk flow and diffusive flow of internal energy, the latter driven by a temperature gradient within the flowing material, and being defined with respect to the local center of mass of the bulk flow. In this case of a virtually closed system, because of the zero matter transfer, as noted above, one can safely distinguish between transfer of energy as work, and transfer of internal energy as heat.

Second Law of Thermodynamics

The second law of thermodynamics states that the total entropy of an isolated system can never decrease over time. The total entropy of a system and its surroundings can remain constant in ideal cases where the system is in thermodynamic equilibrium, or is undergoing a (fictive) reversible process. In all processes that occur, including spontaneous processes, the total entropy of the system and its surroundings increases and the process is irreversible in the thermodynamic sense. The increase in entropy accounts for the irreversibility of natural processes, and the asymmetry between future and past.

Historically, the second law was an empirical finding that was accepted as an axiom of thermodynamic theory. Statistical mechanics, classical or quantum, explains the microscopic origin of the law.

The second law has been expressed in many ways. Its first formulation is credited to the French scientist Sadi Carnot, who in 1824 showed that there is an upper limit to the efficiency of conversion of heat to work, in a heat engine, also referred to as Carnot's rule or limit.

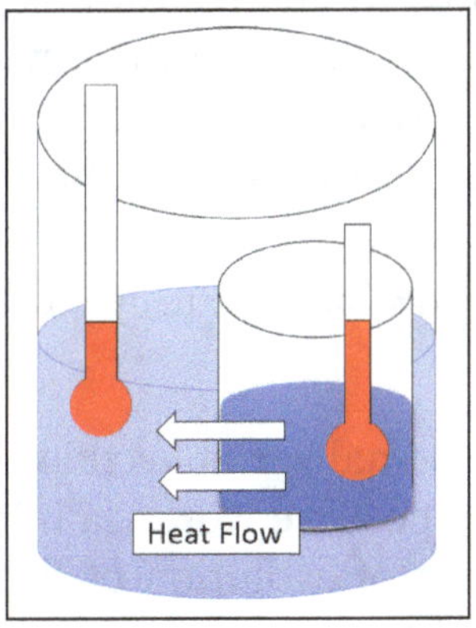

Heat flow from hot water to cold water.

The first law of thermodynamics provides the basic definition of internal energy, associated with all thermodynamic systems, and states the rule of conservation of energy. The second law is concerned with the direction of natural processes. It asserts that a natural process runs only in one sense, and is not reversible. For example, heat always flows spontaneously from hotter to colder bodies, and never the reverse, unless external work is performed on the system. The explanation of the phenomena was given in terms of entropy. Total entropy (S) can never decrease over time for an isolated system because the entropy of an isolated system spontaneously evolves toward thermodynamic equilibrium: the entropy should stay the same or increase.

In a fictive reversible process, an infinitesimal increment in the entropy (dS) of a system is defined to result from an infinitesimal transfer of heat (δQ) to a closed system (which allows the entry or exit of energy – but not mass transfer) divided by the common temperature (T) of the system in equilibrium and the surroundings which supply the heat:

$$\mathrm{d}S = \frac{\delta Q}{T} \qquad \text{(closed system, idealized fictive reversible process).}$$

Different notations are used for infinitesimal amounts of heat (δ) and infinitesimal amounts of entropy (d) because entropy is a function of state, while heat, like work, is not. For an actually possible infinitesimal process without exchange of mass with the surroundings, the second law requires that the increment in system entropy fulfills the inequality:

$$\mathrm{d}S > \frac{\delta Q}{T_{surr}} \qquad \text{(closed system, actually possible, irreversible process).}$$

This is because a general process for this case may include work being done on the system by its surroundings, which can have frictional or viscous effects inside the system, because a chemical reaction may be in progress, or because heat transfer actually occurs only irreversibly, driven by a finite difference between the system temperature (T) and the temperature of the surroundings (T_{surr}). Note that the equality still applies for pure heat flow,

$$S = \frac{\delta Q}{T} \qquad \text{(actually possible quasistatic irreversible process without composition change).}$$

which is the basis of the accurate determination of the absolute entropy of pure substances from measured heat capacity curves and entropy changes at phase transitions, i.e. by calorimetry. Introducing a set of internal variables ξ to describe the deviation of a thermodynamic system in physical equilibrium (with the required well-defined uniform pressure P and temperature T) from the chemical equilibrium state, one can record the equality:

$$\mathrm{d}S = \frac{\delta Q}{T} - \frac{1}{T}\sum_j \Xi_j \delta\xi_j \qquad \text{(closed system, actually possible quasistatic irreversible process).}$$

The second term represents work of internal variables that can be perturbed by external influences, but the system cannot perform any positive work via internal variables. This statement introduces the impossibility of the reversion of evolution of the thermodynamic system in time and can be considered as a formulation of *the second principle of thermodynamics* – the formulation, which is, of course, equivalent to the formulation of the principle in terms of entropy.

The zeroth law of thermodynamics in its usual short statement allows recognition that two bodies in a relation of thermal equilibrium have the same temperature, especially that a test body has the same temperature as a reference thermometric body. For a body in thermal equilibrium with another, there are indefinitely many empirical temperature scales, in general respectively depending on the properties of a particular reference thermometric body. The second law allows a distinguished temperature scale, which defines an absolute, thermodynamic temperature, independent of the properties of any particular reference thermometric body.

Various Statements of the Law

The second law of thermodynamics may be expressed in many specific ways, the most prominent classical statements being the statement by Rudolf Clausius, the statement by Lord Kelvin, and the statement in axiomatic thermodynamics by Constantin Carathéodory. These statements cast the law in general physical terms citing the impossibility of certain processes. The Clausius and the Kelvin statements have been shown to be equivalent.

Carnot's Principle

The historical origin of the second law of thermodynamics was in Carnot's principle. It refers to a cycle of a Carnot heat engine, fictively operated in the limiting mode of extreme slowness known as quasi-static, so that the heat and work transfers are between subsystems that are always in their own internal states of thermodynamic equilibrium. The Carnot engine is an idealized device of special interest to engineers who are concerned with the efficiency of heat engines. Carnot's principle was recognized by Carnot at a time when the caloric theory of heat was seriously considered, before the recognition of the first law of thermodynamics, and before the mathematical expression of the concept of entropy. Interpreted in the light of the first law, it is physically equivalent to the second law of thermodynamics, and remains valid today. In concise terms, Carnot's original statement was made from the viewpoint of the caloric theory, before the discovery of the first law of thermodynamics:

> "The production of motive power is due not to any real consumption of caloric, but to its transport from a warm body to a cold body."

In modern terms, Carnot's principle may be stated more elaborately:

> "The efficiency of a quasi-static or reversible Carnot cycle depends only on the temperatures of the two heat reservoirs, and is the same, whatever the working substance. A Carnot engine operated in this way is the most efficient possible heat engine using those two temperatures."

Clausius Statement

The German scientist Rudolf Clausius laid the foundation for the second law of thermodynamics in 1850 by examining the relation between heat transfer and work. His formulation of the second law, which was published in German in 1854, is known as the *Clausius statement*:

> "Heat can never pass from a colder to a warmer body without some other change, connected therewith, occurring at the same time."

The statement by Clausius uses the concept of 'passage of heat'. As is usual in thermodynamic discussions, this means 'net transfer of energy as heat', and does not refer to contributory transfers one way and the other.

Heat cannot spontaneously flow from cold regions to hot regions without external work being performed on the system, which is evident from ordinary experience of refrigeration, for example. In a refrigerator, heat flows from cold to hot, but only when forced by an external agent, the refrigeration system.

Kelvin Statement

Lord Kelvin expressed the second law as:

> "It is impossible, by means of inanimate material agency, to derive mechanical effect from any portion of matter by cooling it below the temperature of the coldest of the surrounding objects."

Equivalence of the Clausius and the Kelvin Statements

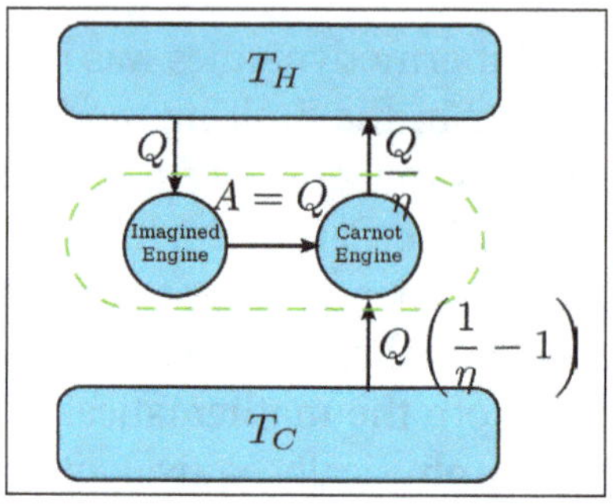

Derive Kelvin Statement from Clausius Statement.

Suppose there is an engine violating the Kelvin statement: i.e., one that drains heat and converts it completely into work in a cyclic fashion without any other result. Now pair it with a reversed Carnot engine as shown by the figure. The net and sole effect of this newly created engine consisting of the two engines mentioned is transferring heat $\ddot{A}Q = Q\left(\frac{1}{\eta} - 1\right)$ from the cooler reservoir to the hotter one, which violates the Clausius statement (This is a consequence of the first law of thermodynamics, as for the total system's energy to remain the same, $\text{Input} + \text{Output} = 0 \Rightarrow Q + \frac{Q}{\eta} = -Q_c$, so therefore $Q_c = Q\left(\frac{1}{\eta} - 1\right)$). Thus a violation of the Kelvin statement implies a violation of the Clausius statement, i.e. the Clausius statement implies the Kelvin statement. We can prove in a similar manner that the Kelvin statement implies the Clausius statement, and hence the two are equivalent.

Planck's Proposition

Planck offered the following proposition as derived directly from experience. This is sometimes regarded as his statement of the second law, but he regarded it as a starting point for the derivation of the second law.

It is impossible to construct an engine which will work in a complete cycle, and produce no effect except the raising of a weight and cooling of a heat reservoir.

Relation between Kelvin's Statement and Planck's Proposition

It is almost customary in textbooks to speak of the "Kelvin-Planck statement" of the law, as for example in the text by ter Haar and Wergeland.

The Kelvin–Planck statement (or the *heat engine statement*) of the second law of thermodynamics states that:

> "It is impossible to devise a cyclically operating device, the sole effect of which is to absorb energy in the form of heat from a single thermal reservoir and to deliver an equivalent amount of work."

Planck's Statement

Planck stated the second law as follows:

> "Every process occurring in nature proceeds in the sense in which the sum of the entropies of all bodies taking part in the process is increased. In the limit, i.e. for reversible processes, the sum of the entropies remains unchanged."

Rather like Planck's statement is that of Uhlenbeck and Ford for *irreversible phenomena*:

> "in an irreversible or spontaneous change from one equilibrium state to another (as for example the equalization of temperature of two bodies A and B, when brought in contact) the entropy always increases."

Principle of Carathéodory

Constantin Carathéodory formulated thermodynamics on a purely mathematical axiomatic foundation. His statement of the second law is known as the Principle of Carathéodory, which may be formulated as follows:

> "In every neighborhood of any state S of an adiabatically enclosed system there are states inaccessible from S."

With this formulation, he described the concept of adiabatic accessibility for the first time and provided the foundation for a new subfield of classical thermodynamics, often called geometrical thermodynamics. It follows from Carathéodory's principle that quantity of energy quasi-statically transferred as heat is a holonomic process function.

Though it is almost customary in textbooks to say that Carathéodory's principle expresses the second law and to treat it as equivalent to the Clausius or to the Kelvin-Planck statements, such is not the case. To get all the content of the second law, Carathéodory's principle needs to be supplemented by Planck's principle, that isochoric work always increases the internal energy of a closed system that was initially in its own internal thermodynamic equilibrium.

Planck's Principle

In 1926, Max Planck wrote an important paper on the basics of thermodynamics. He indicated the principle:

> "The internal energy of a closed system is increased by an adiabatic process, throughout the duration of which, the volume of the system remains constant."

This formulation does not mention heat and does not mention temperature, nor even entropy, and does not necessarily implicitly rely on those concepts, but it implies the content of the second law. A closely related statement is that "Frictional pressure never does positive work." Using a now-obsolete form of words, Planck himself wrote: "The production of heat by friction is irreversible."

Not mentioning entropy, this principle of Planck is stated in physical terms. It is very closely related to the Kelvin statement given just above. It is relevant that for a system at constant volume and mole numbers, the entropy is a monotonic function of the internal energy. Nevertheless, this principle of Planck is not actually Planck's preferred statement of the second law.

A statement that in a sense is complementary to Planck's principle is made by Borgnakke and Sonntag. They do not offer it as a full statement of the second law:

> "there is only one way in which the entropy of a closed system can be decreased, and that is to transfer heat from the system."

Differing from Planck's just foregoing principle, this one is explicitly in terms of entropy change. Removal of matter from a system can also decrease its entropy.

Statement for a System that has a Known Expression of its Internal Energy as a Function of Its Extensive State Variables: The second law has been shown to be equivalent to the internal energy U being a weakly convex function, when written as a function of extensive properties (mass, volume, entropy).

Corollaries

Perpetual Motion of the Second Kind

Before the establishment of the second law, many people who were interested in inventing a perpetual motion machine had tried to circumvent the restrictions of first law of thermodynamics by extracting the massive internal energy of the environment as the power of the machine. Such a machine is called a "perpetual motion machine of the second kind". The second law declared the impossibility of such machines.

Carnot Theorem

Carnot's theorem is a principle that limits the maximum efficiency for any possible engine. The efficiency solely depends on the temperature difference between the hot and cold thermal reservoirs. Carnot's theorem states:

- All irreversible heat engines between two heat reservoirs are less efficient than a Carnot engine operating between the same reservoirs.
- All reversible heat engines between two heat reservoirs are equally efficient with a Carnot engine operating between the same reservoirs.

In his ideal model, the heat of caloric converted into work could be reinstated by reversing the motion of the cycle, a concept subsequently known as thermodynamic reversibility. Carnot, however, further postulated that some caloric is lost, not being converted to mechanical work. Hence, no real heat engine could realise the Carnot cycle's reversibility and was condemned to be less efficient.

Though formulated in terms of caloric, rather than entropy, this was an early insight into the second law.

Clausius Inequality

The Clausius theorem states that in a cyclic process:

$$\oint \frac{\delta Q}{T} \leq 0.$$

The equality holds in the reversible case and the strict inequality holds in the irreversible case. The reversible case is used to introduce the state function entropy. This is because in cyclic processes the variation of a state function is zero from state functionality.

Thermodynamic Temperature

For an arbitrary heat engine, the efficiency is:

$$\eta = \frac{W_n}{q_H} = \frac{q_H - q_C}{q_H} = 1 - \frac{q_C}{q_H}$$

where W_n is for the net work done per cycle. Thus the efficiency depends only on q_C/q_H.

Carnot's theorem states that all reversible engines operating between the same heat reservoirs are equally efficient. Thus, any reversible heat engine operating between temperatures T_1 and T_2 must have the same efficiency, that is to say, the efficiency is the function of temperatures only:

$$\frac{q_C}{q_H} = f(T_H, T_C)$$

In addition, a reversible heat engine operating between temperatures T_1 and T_3 must have the same efficiency as one consisting of two cycles, one between T_1 and another (intermediate) temperature T_2, and the second between T_2 andT_3. This can only be the case if:

$$f(T_1, T_3) = \frac{q_3}{q_1} = \frac{q_2 q_3}{q_1 q_2} = f(T_1, T_2) f(T_2, T_3).$$

Now consider the case where T_1 is a fixed reference temperature: the temperature of the triple point of water. Then for any T_2 and T_3:

$$f(T_2, T_3) = \frac{f(T_1, T_3)}{f(T_1, T_2)} = \frac{273.16 \cdot f(T_1, T_3)}{273.16 \cdot f(T_1, T_2)}.$$

Therefore, if thermodynamic temperature is defined by:

$$T = 273.16 \cdot f(T_1, T)$$

then the function *f*, viewed as a function of thermodynamic temperature, is simply:

$$f(T_2, T_3) = \frac{T_3}{T_2},$$

and the reference temperature T_1 will have the value 273.16. (Any reference temperature and any positive numerical value could be used – the choice here corresponds to the Kelvin scale.)

Entropy

According to the Clausius equality, for a reversible process:

$$\oint \frac{\delta Q}{T} = 0$$

That means the line integral $\int_L \frac{\delta Q}{T}$ is path independent for reversible processes.

So we can define a state function S called entropy, which for a reversible process or for pure heat transfer satisfies:

$$dS = \frac{\delta Q}{T}$$

With this we can only obtain the difference of entropy by integrating the above formula. To obtain the absolute value, we need the third law of thermodynamics, which states that $S = 0$ at absolute zero for perfect crystals.

For any irreversible process, since entropy is a state function, we can always connect the initial and terminal states with an imaginary reversible process and integrating on that path to calculate the difference in entropy.

Now reverse the reversible process and combine it with the said irreversible process. Applying the Clausius inequality on this loop:

$$-\Delta S + \int \frac{\delta Q}{T} = \oint \frac{\delta Q}{T} < 0$$

Thus,

$$\Delta S \geq \int \frac{\delta Q}{T}$$

where the equality holds if the transformation is reversible.

Notice that if the process is an adiabatic process, then $\delta Q = 0$, so $\Delta S \geq 0$.

Energy and Available useful Work

An important and revealing idealized special case is to consider applying the Second Law to the scenario of an isolated system (called the total system or universe), made up of two parts: a sub-system of interest, and the sub-system's surroundings. These surroundings are imagined to be so large that they can be considered as an *unlimited* heat reservoir at temperature T_R and pressure P_R – so that no matter how much heat is transferred to (or from) the sub-system, the temperature of the surroundings will remain T_R; and no matter how much the volume of the sub-system expands (or contracts), the pressure of the surroundings will remain P_R.

Whatever changes to dS and dS_R occur in the entropies of the sub-system and the surroundings individually, according to the Second Law the entropy S_{tot} of the isolated total system must not decrease:

$$dS_{\text{tot}} = dS + dS_R \geq 0$$

According to the first law of thermodynamics, the change dU in the internal energy of the sub-system is the sum of the heat δq added to the sub-system, *less* any work δw done *by* the sub-system, *plus* any net chemical energy entering the sub-system $d\sum\mu_{iR}N_i$, so that:

$$dU = \delta q - \delta w + d(\sum \mu_{iR} N_i)$$

where μ_{iR} are the chemical potentials of chemical species in the external surroundings.

Now the heat leaving the reservoir and entering the sub-system is:

$$\delta q = T_R(-dS_R) \leq T_R dS$$

where we have first used the definition of entropy in classical thermodynamics (alternatively, in statistical thermodynamics, the relation between entropy change, temperature and absorbed heat can be derived); and then the Second Law inequality from above.

It therefore follows that any net work δw done by the sub-system must obey:

$$\delta w \leq -dU + T_R dS + \sum \mu_{iR} dN_i$$

It is useful to separate the work δw done by the subsystem into the *useful* work δw_u that can be done *by* the sub-system, over and beyond the work $p_R\, dV$ done merely by the sub-system expanding against the surrounding external pressure, giving the following relation for the useful work (exergy) that can be done:

$$\delta w_u \leq -d(U - T_R S + p_R V - \sum \mu_{iR} N_i)$$

It is convenient to define the right-hand-side as the exact derivative of a thermodynamic potential, called the *availability* or *exergy E* of the subsystem:

$$E = U - T_R S + p_R V - \sum \mu_{iR} N_i$$

The Second Law therefore implies that for any process which can be considered as divided simply into a subsystem, and an unlimited temperature and pressure reservoir with which it is in contact:

$$dE + \delta w_u \leq 0$$

i.e. the change in the subsystem's exergy plus the useful work done *by* the subsystem (or, the change in the subsystem's exergy less any work, additional to that done by the pressure reservoir, done *on* the system) must be less than or equal to zero.

In sum, if a proper *infinite-reservoir-like* reference state is chosen as the system surroundings in the real world, then the Second Law predicts a decrease in E for an irreversible process and no change for a reversible process.

$$dS_{tot} \geq 0 \text{ is equivalent to } dE + \delta w_u \leq 0$$

This expression together with the associated reference state permits a design engineer working at the macroscopic scale (above the thermodynamic limit) to utilize the Second Law without directly measuring or considering entropy change in a total isolated system. Those changes have already been considered by the assumption that the system under consideration can reach equilibrium with the reference state without altering the reference state. An efficiency for a process or collection of processes that compares it to the reversible ideal may also be found.

This approach to the Second Law is widely utilized in engineering practice, environmental accounting, systems ecology, and other disciplines.

The Second Law in Chemical Thermodynamics

For a spontaneous chemical process in a closed system at constant temperature and pressure without non-*PV* work, the Clausius inequality $\Delta S > Q/T_{surr}$ transforms into a condition for the change in Gibbs free energy:

$$\Delta G < 0$$

or $dG < 0$. For a similar process at constant temperature and volume, the change in Helmholtz free energy must be negative, $\Delta A < 0$. Thus, a negative value of the change in free energy (G or A) is a necessary condition for a process to be spontaneous. This is the most useful form of the second law of thermodynamics in chemistry, where free-energy changes can be calculated from tabulated enthalpies of formation and standard molar entropies of reactants and products. The chemical equilibrium condition at constant T and p without electrical work is $dG = 0$.

Statistical Mechanics

Statistical mechanics gives an explanation for the second law by postulating that a material is composed of atoms and molecules which are in constant motion. A particular set of positions and velocities for each particle in the system is called a microstate of the system and because of the constant motion, the system is constantly changing its microstate. Statistical mechanics postulates that, in equilibrium, each microstate that the system might be in is equally likely to occur, and when this assumption is made, it leads directly to the conclusion that the second law must hold in a statistical sense. That is, the second law will hold on average, with a statistical variation on the order of $1/\sqrt{N}$ where N is the number of particles in the system. For everyday (macroscopic) situations, the probability that the second law will be violated is practically zero. However, for systems with a small number of particles, thermodynamic parameters, including the entropy, may show significant statistical deviations from that predicted by the second law. Classical thermodynamic theory does not deal with these statistical variations.

Derivation from Statistical Mechanics

The first mechanical argument of the Kinetic theory of gases that molecular collisions entail an equalization of temperatures and hence a tendency towards equilibrium was due to James Clerk Maxwell in 1860; Ludwig Boltzmann with his H-theorem of 1872 also argued that due to collisions gases should over time tend toward the Maxwell-Boltzmann distribution.

Due to Loschmidt's paradox, derivations of the Second Law have to make an assumption regarding the past, namely that the system is uncorrelated at some time in the past; this allows for simple probabilistic treatment. This assumption is usually thought as a boundary condition, and thus the second Law is ultimately a consequence of the initial conditions somewhere in the past, probably at the beginning of the universe (the Big Bang), though other scenarios have also been suggested.

Given these assumptions, in statistical mechanics, the Second Law is not a postulate, rather it is a consequence of the fundamental postulate, also known as the equal prior probability postulate, so long as one is clear that simple probability arguments are applied only to the future, while for the past there are auxiliary sources of information which tell us that it was low entropy. The first part of the second law, which states that the entropy of a thermally isolated system can only increase, is a trivial consequence of the equal prior probability postulate, if we restrict the notion of the entropy to systems in thermal equilibrium. The entropy of an isolated system in thermal equilibrium containing an amount of energy of E is:

$$S = k_{\mathrm{B}} \ln\left[\Omega(E)\right]$$

where $\Omega(E)$ is the number of quantum states in a small interval between E and $E+\delta E$. Here δE is a macroscopically small energy interval that is kept fixed. Strictly speaking this means that the entropy depends on the choice of δE. However, in the thermodynamic limit (i.e. in the limit of infinitely large system size), the specific entropy (entropy per unit volume or per unit mass) does not depend on δE.

Suppose we have an isolated system whose macroscopic state is specified by a number of variables. These macroscopic variables can, e.g, refer to the total volume, the positions of pistons in the system, etc. Then Ω will depend on the values of these variables. If a variable is not fixed, (e.g. we do not clamp a piston in a certain position), then because all the accessible states are equally likely in equilibrium, the free variable in equilibrium will be such that Ω is maximized as that is the most probable situation in equilibrium.

If the variable was initially fixed to some value then upon release and when the new equilibrium has been reached, the fact the variable will adjust itself so that Ω is maximized, implies that the entropy will have increased or it will have stayed the same (if the value at which the variable was fixed happened to be the equilibrium value). Suppose we start from an equilibrium situation and we suddenly remove a constraint on a variable. Then right after we do this, there are a number Ω of accessible microstates, but equilibrium has not yet been reached, so the actual probabilities of the system being in some accessible state are not yet equal to the prior probability of $1/\Omega$. We have already seen that in the final equilibrium state, the entropy will have increased or have stayed the same relative to the previous equilibrium state. Boltzmann's H-theorem, however, proves that the quantity H increases monotonically as a function of time during the intermediate out of equilibrium state.

Derivation of the Entropy Change for Reversible Processes

The second part of the Second Law states that the entropy change of a system undergoing a reversible process is given by:

$$dS = \frac{\delta Q}{T}$$

where the temperature is defined as:

$$\frac{1}{k_B T} \equiv \beta \equiv \frac{d \ln\left[\Omega(E)\right]}{dE}$$

For the justification for this definition. Suppose that the system has some external parameter, x, that can be changed. In general, the energy eigenstates of the system will depend on x. According to the adiabatic theorem of quantum mechanics, in the limit of an infinitely slow change of the system's Hamiltonian, the system will stay in the same energy eigenstate and thus change its energy according to the change in energy of the energy eigenstate it is in.

The generalized force, X, corresponding to the external variable x is defined such that Xdx is the work performed by the system if x is increased by an amount dx. E.g., if x is the volume, then X is the pressure. The generalized force for a system known to be in energy eigenstate E_r is given by:

$$X = -\frac{dE_r}{dx}$$

Since the system can be in any energy eigenstate within an interval of δE, we define the generalized force for the system as the expectation value of the above expression:

$$X = -\left\langle \frac{dE_r}{dx} \right\rangle$$

To evaluate the average, we partition the $\Omega(E)$ energy eigenstates by counting how many of them have a value for $\frac{dE_r}{dx}$ within a range between Y and $Y + \delta Y$. Calling this number $\Omega_Y(E)$, we have:

$$\Omega(E) = \sum_Y \Omega_Y(E)$$

The average defining the generalized force can now be written:

$$X = -\frac{1}{\Omega(E)}\sum_{Y} Y\Omega_Y(E)$$

We can relate this to the derivative of the entropy with respect to x at constant energy E as follows. Suppose we change x to x + dx. Then $\Omega(E)$ will change because the energy eigenstates depend on x, causing energy eigenstates to move into or out of the range between E and $E+\delta E$. Let's focus again on the energy eigenstates for which $\frac{dE_r}{dx}$ lies within the range between Y and $Y+\delta Y$. Since these energy eigenstates increase in energy by Y dx, all such energy eigenstates that are in the interval ranging from E – Y dx to E move from below E to above E. There are:

$$N_Y(E) = \frac{\Omega_Y(E)}{\delta E} Y\,dx$$

such energy eigenstates. If $Ydx \leq \delta E$, all these energy eigenstates will move into the range between E and $E+\delta E$ and contribute to an increase in Ω. The number of energy eigenstates that move from below $E+\delta E$ to above $E+\delta E$ is given by $N_Y(E+\delta E)$. The difference:

$$N_Y(E) - N_Y(E+\delta E)$$

is thus the net contribution to the increase in Ω. Note that if Y dx is larger than δE there will be the energy eigenstates that move from below E to above $E+\delta E$. They are counted in both $N_Y(E)$ and $N_Y(E+\delta E)$, therefore the above expression is also valid in that case.

Expressing the above expression as a derivative with respect to E and summing over Y yields the expression:

$$\left(\frac{\partial\Omega}{\partial x}\right)_E = -\sum_Y Y\left(\frac{\partial\Omega_Y}{\partial E}\right)_x = \left(\frac{\partial(\Omega X)}{\partial E}\right)_x$$

The logarithmic derivative of Ω with respect to x is thus given by:

$$\left(\frac{\partial \ln(\Omega)}{\partial x}\right)_E = \beta X + \left(\frac{\partial X}{\partial E}\right)_x$$

The first term is intensive, i.e. it does not scale with system size. In contrast, the last term scales as the inverse system size and will thus vanishes in the thermodynamic limit. We have thus found that:

$$\left(\frac{\partial S}{\partial x}\right)_E = \frac{X}{T}$$

Combining this with:

$$\left(\frac{\partial S}{\partial E}\right)_x = \frac{1}{T}$$

Gives:

$$dS = \left(\frac{\partial S}{\partial E}\right)_x dE + \left(\frac{\partial S}{\partial x}\right)_E dx = \frac{dE}{T} + \frac{X}{T}dx = \frac{\delta Q}{T}$$

Derivation for Systems Described by the Canonical Ensemble

If a system is in thermal contact with a heat bath at some temperature T then, in equilibrium, the probability distribution over the energy eigenvalues are given by the canonical ensemble:

$$P_j = \frac{\exp\left(-\frac{E_j}{k_B T}\right)}{Z}$$

Here Z is a factor that normalizes the sum of all the probabilities to 1, this function is known as the partition function. We now consider an infinitesimal reversible change in the temperature and in the external parameters on which the energy levels depend. It follows from the general formula for the entropy:

$$S = -k_B \sum_j P_j \ln(P_j)$$

that

$$dS = -k_B \sum_j \ln(P_j) dP_j$$

Inserting the formula for P_j for the canonical ensemble in here gives:

$$dS = \frac{1}{T}\sum_j E_j dP_j = \frac{1}{T}\sum_j d(E_j P_j) - \frac{1}{T}\sum_j P_j dE_j = \frac{dE + \delta W}{T} = \frac{\delta Q}{T}$$

Living Organisms

There are two principal ways of formulating thermodynamics, (a) through passages from one state of thermodynamic equilibrium to another, and (b) through cyclic processes, by which the system is left unchanged, while the total entropy of the surroundings is increased. These two ways help to understand the processes of life.

To a fair approximation, living organisms may be considered as examples of (b). Approximately, an animal's physical state cycles by the day, leaving the animal nearly unchanged. Animals take in food, water, and oxygen, and, as a result of metabolism, give out breakdown products and heat. Plants take in radiative energy from the sun, which may be regarded as heat, and carbon dioxide and water. They give out oxygen. In this way they grow. Eventually they die, and their remains rot away, turning mostly back into carbon dioxide and water. This can be regarded as a cyclic process. Overall, the sunlight is from a high temperature source, the sun, and its energy is passed to a lower temperature sink, i.e. radiated into space. This is an increase of entropy of the surroundings of the plant. Thus animals and plants obey the second law of thermodynamics, considered in terms of cyclic processes. Simple concepts of efficiency of heat engines are hardly applicable to this problem because they assume closed systems.

From the thermodynamic viewpoint that considers (a), passages from one equilibrium state to another, only a roughly approximate picture appears, because living organisms are never in states of thermodynamic equilibrium. Living organisms must often be considered as open systems, because they take in nutrients and give out waste products. Thermodynamics of open systems is currently often considered in terms of passages from one state of thermodynamic equilibrium to another, or in terms of flows in the approximation of local thermodynamic equilibrium. The problem for living organisms may be further simplified by the approximation of assuming a steady state with unchanging flows. General principles of entropy production for such approximations are subject to unsettled current debate or research. Nevertheless, ideas derived from this viewpoint on the second law of thermodynamics are enlightening about living creatures.

Gravitational Systems

In systems that do not require for their descriptions the general theory of relativity, bodies always have positive heat capacity, meaning that the temperature rises with energy. Therefore, when energy flows from a high-temperature object to a low-temperature object, the source temperature is decreased while the sink temperature is increased; hence temperature differences tend to diminish over time. This is not always the case for systems in which the gravitational force is important and the general theory of relativity is required. Such systems can spontaneously change towards uneven spread of mass and energy. This applies to the universe in large scale, and consequently it may be difficult or impossible to apply the second law to it.

Non-equilibrium States

The theory of classical or equilibrium thermodynamics is idealized. A main postulate or assumption, often not even explicitly stated, is the existence of systems in their own internal states of thermodynamic equilibrium. In general, a region of space containing a physical system at a given time, that may be found in nature, is not in thermodynamic equilibrium, read in the most stringent terms. In looser terms, nothing in the entire universe is or has ever been truly in exact thermodynamic equilibrium.

For purposes of physical analysis, it is often enough convenient to make an assumption of thermodynamic equilibrium. Such an assumption may rely on trial and error for its justification. If the assumption is justified, it can often be very valuable and useful because it makes available the theory of thermodynamics. Elements of the equilibrium assumption are that a system is observed to be unchanging over an indefinitely long time, and that there are so many particles in a system, that its particulate nature can be entirely ignored. Under such an equilibrium assumption, in general, there are no macroscopically detectable fluctuations. There is an exception, the case of critical states, which exhibit to the naked eye the phenomenon of critical opalescence. For laboratory studies of critical states, exceptionally long observation times are needed.

In all cases, the assumption of thermodynamic equilibrium, once made, implies as a consequence that no putative candidate "fluctuation" alters the entropy of the system.

It can easily happen that a physical system exhibits internal macroscopic changes that are fast enough to invalidate the assumption of the constancy of the entropy. Or that a physical system has so few particles that the particulate nature is manifest in observable fluctuations. Then the assumption of thermodynamic equilibrium is to be abandoned. There is no unqualified general definition of entropy for non-equilibrium states.

There are intermediate cases, in which the assumption of local thermodynamic equilibrium is a very good approximation, but strictly speaking it is still an approximation, not theoretically ideal. For non-equilibrium situations in general, it may be useful to consider statistical mechanical definitions of other quantities that may be conveniently called 'entropy', but they should not be confused or conflated with thermodynamic entropy properly defined for the second law. These other quantities indeed belong to statistical mechanics, not to thermodynamics, the primary realm of the second law.

Even though the applicability of the second law of thermodynamics is limited for non-equilibrium systems, the laws governing such systems are still being discussed. One of the guiding principles for systems which are far from equilibrium is the maximum entropy production principle. It states that a system away from equilibrium evolves in such a way as to maximize entropy production, given present constraints.

Arrow of Time

The second law of thermodynamics is a physical law that is not symmetric to reversal of the time direction. This does not conflict with notions that have been observed of the fundamental laws of physics, namely CPT symmetry, since the second law applies statistically, it is hypothesized, on time-asymmetric *boundary conditions.*

The second law has been proposed to supply a partial explanation of the difference between moving forward and backwards in time, such as why the cause precedes the effect (the causal arrow of time).

Irreversibility

Irreversibility in thermodynamic processes is a consequence of the asymmetric character of thermodynamic operations, and not of any internally irreversible microscopic properties of the bodies. Thermodynamic operations are macroscopic external interventions imposed on the participating bodies, not derived from their internal properties. There are reputed "paradoxes" that arise from failure to recognize this.

Loschmidt's Paradox

Loschmidt's paradox, also known as the reversibility paradox, is the objection that it should not be possible to deduce an irreversible process from the time-symmetric dynamics that describe the microscopic evolution of a macroscopic system.

In the opinion of Schrödinger, "It is now quite obvious in what manner you have to reformulate the law of entropy – or for that matter, all other irreversible statements – so that they be capable of being derived from reversible models. You must not speak of one isolated system but at least of two, which you may for the moment consider isolated from the rest of the world, but not always from each other." The two systems are isolated from each other by the wall, until it is removed by the thermodynamic operation, as envisaged by the law. The thermodynamic operation is externally imposed, not subject to the reversible microscopic dynamical laws that govern the constituents of the systems. It is the cause of the irreversibility. The statement of the law complies with Schrödinger's advice. The cause–effect relation is logically prior to the second law, not derived from it.

Poincaré Recurrence Theorem

The Poincaré recurrence theorem considers a theoretical microscopic description of an isolated physical system. This may be considered as a model of a thermodynamic system after a thermodynamic operation has removed an internal wall. The system will, after a sufficiently long time, return to a microscopically defined state very close to the initial one. The Poincaré recurrence time is the length of time elapsed until the return. It is exceedingly long, likely longer than the life of the universe, and depends sensitively on the geometry of the wall that was removed by the thermodynamic operation. The recurrence theorem may be perceived as apparently contradicting the second law of thermodynamics. More obviously, however, it is simply a microscopic model of thermodynamic equilibrium in an isolated system formed by removal of a wall between two systems. For a typical thermodynamical system, the recurrence time is so large (many many times longer than the lifetime of the universe) that, for all practical purposes, one cannot observe the recurrence. One might wish, nevertheless, to imagine that one could wait for the Poincaré recurrence, and then re-insert the wall that was removed by the thermodynamic operation. It is then evident that the appearance of irreversibility is due to the utter unpredictability of the Poincaré recurrence given only that the initial state was one of thermodynamic equilibrium, as is the case in macroscopic thermodynamics. Even if one could wait for it, one has no practical possibility of picking the right instant at which to re-insert the wall. The Poincaré recurrence theorem provides a solution to Loschmidt's paradox. If an isolated thermodynamic system could be monitored over increasingly many multiples of the average Poincaré recurrence time, the thermodynamic behavior of the system would become invariant under time reversal.

Maxwell's Demon

James Clerk Maxwell imagined one container divided into two parts, A and B. Both parts are filled with the same gas at equal temperatures and placed next to each other, separated by a wall. Observing the molecules on both sides, an imaginary demon guards a microscopic trapdoor in the wall. When a faster-than-average molecule from A flies towards the trapdoor, the demon opens it, and the molecule will fly from A to B. The average speed of the molecules in B will have increased while in A they will have slowed down on average. Since average molecular speed corresponds to temperature, the temperature decreases in A and increases in B, contrary to the second law of thermodynamics.

One response to this question was suggested in 1929 by Leó Szilárd and later by Léon Brillouin. Szilárd pointed out that a real-life Maxwell's demon would need to have some means of measuring molecular speed, and that the act of acquiring information would require an expenditure of energy.

Maxwell's 'demon' repeatedly alters the permeability of the wall between A and B. It is therefore performing thermodynamic operations on a microscopic scale, not just observing ordinary spontaneous or natural macroscopic thermodynamic processes.

Third Law of Thermodynamics

The Third law of thermodynamics is sometimes stated as follows, regarding the properties of closed systems in thermodynamic equilibrium:

> "The Entropy of a system approaches a constant value as its temperature approaches absolute zero."

This constant value cannot depend on any other parameters characterizing the closed system, such as pressure or applied magnetic field. At absolute zero (zero kelvin) the system must be in a state with the minimum possible energy. Entropy is related to the number of accessible microstates, and there is typically one unique state (called the ground state) with minimum energy. In such a case, the entropy at absolute zero will be exactly zero. If the system does not have a well-defined order (if its order is glassy, for example), then there may remain some finite entropy as the system is brought to very low temperatures, either because the system becomes locked into a configuration with non-minimal energy or because the minimum energy state is non-unique. The constant value is called the residual entropy of the system. The entropy is essentially a state-function meaning the inherent value of different atoms, molecules, and other configurations of particles including subatomic or atomic material is defined by entropy, which can be discovered near 0 K. The Nernst–Simon statement of the third law of thermodynamics concerns thermodynamic processes at a fixed, low temperature:

> "The entropy change associated with any condensed system undergoing a reversible isothermal process approaches zero as the temperature at which it is performed approaches 0 K."

Here a condensed system refers to liquids and solids. A classical formulation by Nernst (actually a consequence of the Third Law) is:

> "It is impossible for any process, no matter how idealized, to reduce the entropy of a system to its absolute-zero value in a finite number of operations."

There also exists a formulation of the Third Law which approaches the subject by postulating a specific energy behavior:

> "If the composite of two thermodynamic systems constitutes an isolated system, then any energy exchange in any form between those two systems is bounded."

The third law was developed by chemist Walther Nernst during the years 1906–12, and is therefore often referred to as Nernst's theorem or Nernst's postulate. The third law of thermodynamics states that the entropy of a system at absolute zero is a well-defined constant. This is because a system at zero temperature exists in its ground state, so that its entropy is determined only by the degeneracy of the ground state.

In 1912 Nernst stated the law thus: "It is impossible for any procedure to lead to the isotherm $T = 0$ in a finite number of steps."

An alternative version of the third law of thermodynamics as stated by Gilbert N. Lewis and Merle Randall in 1923:

If the entropy of each element in some (perfect) crystalline state be taken as zero at the absolute zero of temperature, every substance has a finite positive entropy; but at the absolute zero of temperature the entropy may become zero, and does so become in the case of perfect crystalline substances.

This version states not only ΔS will reach zero at 0 K, but S itself will also reach zero as long as the crystal has a ground state with only one configuration. Some crystals form defects which cause a residual entropy. This residual entropy disappears when the kinetic barriers to transitioning to one ground state are overcome.

With the development of statistical mechanics, the third law of thermodynamics (like the other laws) changed from a *fundamental* law (justified by experiments) to a *derived* law (derived from even more basic laws). The basic law from which it is primarily derived is the statistical-mechanics definition of entropy for a large system:

$$S - S_0 = k_B \ln \Omega$$

where S is entropy, k_B is the Boltzmann constant, and Ω is the number of microstates consistent with the macroscopic configuration. The counting of states is from the reference state of absolute zero, which corresponds to the entropy of S_0.

In simple terms, the third law states that the entropy of a perfect crystal of a pure substance approaches zero as the temperature approaches zero. The alignment of a perfect crystal leaves no ambiguity as to the location and orientation of each part of the crystal. As the energy of the crystal is reduced, the vibrations of the individual atoms are reduced to nothing, and the crystal becomes the same everywhere.

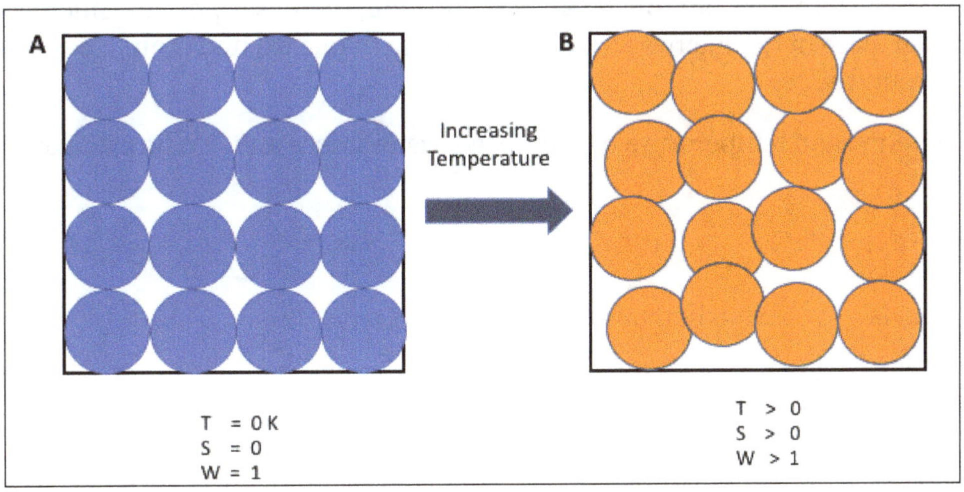

a) Single possible configuration for a system at absolute zero, i.e., only one microstate is accessible. Thus S = k ln W = 0. b) At temperatures greater than absolute zero, multiple microstates are accessible due to atomic vibration (exaggerated in the figure). Since the number of accessible microstates is greater than 1, S = k ln W > 0.

The third law provides an absolute reference point for the determination of entropy at any other temperature. The entropy of a closed system, determined relative to this zero point, is then the *absolute* entropy of that system. Mathematically, the absolute entropy of any system at zero temperature is the natural log of the number of ground states times Boltzmann's constant $k_B = 1.38 \times 10^{-23}$ J K^{-1}.

The entropy of a *perfect* crystal lattice as defined by Nernst's theorem is zero provided that its ground state is unique, because ln(1) = 0. If the system is composed of one-billion atoms, all alike, and lie within the matrix of a perfect crystal, the number of combinations of one-billion identical things taken one-billion at a time is Ω = 1. Hence:

$$S - S_0 = k_B \ln \Omega = k_B \ln 1 = 0$$

The difference is zero, hence the initial entropy S_0 can be any selected value so long as all other such calculations include that as the initial entropy. As a result, the initial entropy value of zero is selected $S_0 = 0$ is used for convenience.

$$S - S_0 = S - 0 = 0$$

$$S = 0$$

Entropy Change of a Crystal Lattice Heated by an Incoming Photon

Suppose a system consisting of a crystal lattice with volume V of N identical atoms at T= 0 K, and an incoming photon of wavelength λ and energy ε.

Initially, there is only one accessible microstate:

$$S_0 = k_B \ln \Omega = k_B \ln 1 = 0$$

Let's assume the crystal lattice absorbs the incoming photon. There is a unique atom in the lattice that interacts and absorbs this photon. So after absorption, there is N possible microstates accessible by the system, each of the microstates corresponding to one excited atom, and the other atoms remaining at ground state.

The entropy, energy, and temperature of the closed system rises and can be calculated. The entropy change is:

$$\Delta S = S - S_0 = k_B \ln \Omega$$

From the second law of thermodynamics:

$$\Delta S = S - S_0 = \frac{\delta Q}{T}$$

Hence:

$$\Delta S = S - S_0 = k_B \ln(\Omega) = \frac{\delta Q}{T}$$

Calculating entropy change:

$$S - 0 = k_B \ln N = 1.38 \times 10^{-23} \times \ln(3 \times 10^{22}) = 70 \times 10^{-23}\,\text{JK}^{-1}$$

We assume N = 3.10[22] and λ = 1 cm. The energy change of the system as a result of absorbing the single photon whose energy is ε:

$$\delta Q = \epsilon = \frac{hc}{\lambda} = \frac{6.62\times10^{-34}\,\mathrm{J\cdot s}\times3\times10^{8}\,\mathrm{ms}^{-1}}{0.01\mathrm{m}} = 2\times10^{-23}\,\mathrm{J}$$

The temperature of the closed system rises by:

$$T = \frac{\epsilon}{\Delta S} = \frac{2\times10^{-23}\,\mathrm{J}}{70\times10^{-23}\,\mathrm{JK}^{-1}} = 0.02857K$$

This can be interpreted as the average temperature of the system over the range from $0 < S < 70\times10^{-23}\,\mathrm{JK}^{-1}$. A single atom was assumed to absorb the photon but the temperature and entropy change characterizes the entire system.

Systems with Non-zero Entropy at Absolute Zero

An example of a system which does not have a unique ground state is one whose net spin is a half-integer, for which time-reversal symmetry gives two degenerate ground states. For such systems, the entropy at zero temperature is at least k_B*ln(2) (which is negligible on a macroscopic scale). Some crystalline systems exhibit geometrical frustration, where the structure of the crystal lattice prevents the emergence of a unique ground state. Ground-state helium (unless under pressure) remains liquid.

In addition, glasses and solid solutions retain large entropy at 0 K, because they are large collections of nearly degenerate states, in which they become trapped out of equilibrium. Another example of a solid with many nearly-degenerate ground states, trapped out of equilibrium, is ice Ih, which has "proton disorder".

For the entropy at absolute zero to be zero, the magnetic moments of a perfectly ordered crystal must themselves be perfectly ordered; from an entropic perspective, this can be considered to be part of the definition of a "perfect crystal". Only ferromagnetic, antiferromagnetic, and diamagnetic materials can satisfy this condition. However, ferromagnetic materials do not, in fact, have zero entropy at zero temperature, because the spins of the unpaired electrons are all aligned and this gives a ground-state spin degeneracy. Materials that remain paramagnetic at 0 K, by contrast, may have many nearly-degenerate ground states (for example, in a spin glass), or may retain dynamic disorder (a quantum spin liquid).

Consequences

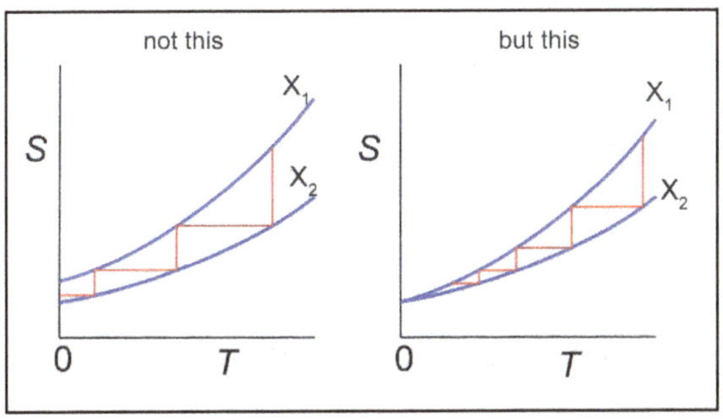

Left side: Absolute zero can be reached in a finite number of steps if $S(0,X_1)\neq S(0,X_2)$.
Right: An infinite number of steps is needed since $S(0,X_1)= S(0,X_2)$.

Absolute Zero

The third law is equivalent to the statement that:

> "It is impossible by any procedure, no matter how idealized, to reduce the temperature of any closed system to zero temperature in a finite number of finite operations."

The reason that $T = 0$ cannot be reached according to the third law is explained as follows: Suppose that the temperature of a substance can be reduced in an isentropic process by changing the parameter X from X_2 to X_1. One can think of a multistage nuclear demagnetization setup where a magnetic field is switched on and off in a controlled way. If there were an entropy difference at absolute zero, $T = 0$ could be reached in a finite number of steps. However, at $T = 0$ there is no entropy difference so an infinite number of steps would be needed. The process is illustrated in figure.

Specific Heat

A non-quantitative description of his third law that Nernst gave at the very beginning was simply that the specific heat can always be made zero by cooling the material down far enough. A modern, quantitative analysis follows.

Supposed that the heat capacity of a sample in the low temperature region has the form of a power law $C(T,X)=C_0T^\alpha$ asymptotically as $T\to 0$, and we wish to find which values of α are compatible with the third law. We have:

$$\int_{T_0}^{T} \frac{C(T',X)}{T'} dT' = \frac{C_0}{\alpha}(T^\alpha - T_0^\alpha).$$

By the discussion of third law, this integral must be bounded as $T_0\to 0$, which is only possible if $\alpha>0$. So the heat capacity must go to zero at absolute zero:

$$\lim_{T\to 0} C(T,X) = 0.$$

if it has the form of a power law. The same argument shows that it cannot be bounded below by a positive constant, even if we drop the power-law assumption.

On the other hand, the molar specific heat at constant volume of a monatomic classical ideal gas, such as helium at room temperature, is given by $C_V=(3/2)R$ with R the molar ideal gas constant. But clearly a constant heat capacity does not satisfy $\lim_{T\to 0} C(T,X) = 0$. That is, a gas with a constant heat capacity all the way to absolute zero violates the third law of thermodynamics. We can verify this more fundamentally by substituting C_V in $C_V = \frac{\pi^2}{2} R \frac{T}{T_F}$, which yields:

$$S(T,V) = S(T_0,V) + \frac{3}{2} R \ln \frac{T}{T_0}$$

In the limit $T_0 \to 0$ this expression diverges, again contradicting the third law of thermodynamics.

The conflict is resolved as follows: At a certain temperature the quantum nature of matter starts to dominate the behavior. Fermi particles follow Fermi–Dirac statistics and Bose particles follow

Bose–Einstein statistics. In both cases the heat capacity at low temperatures is no longer temperature independent, even for ideal gases. For Fermi gases:

$$C_V = \frac{\pi^2}{2} R \frac{T}{T_F}$$

with the Fermi temperature T_F given by:

$$T_F = \frac{1}{8\pi^2} \frac{N_A^2 h^2}{MR} \left(\frac{3\pi^2 N_A}{V_m} \right)^{2/3}$$

Here N_A is Avogadro's number, V_m the molar volume, and M the molar mass.

For Bose gases:

$$C_V = 1.93..R \left(\frac{T}{T_B} \right)^{3/2}$$

with T_B given by:

$$T_B = \frac{1}{11.9..} \frac{N_A^2 h^2}{MR} \left(\frac{N_A}{V_m} \right)^{2/3}$$

The specific heats given by $C_V = \frac{\pi^2}{2} R \frac{T}{T_F}$ and $C_V = 1.93..R \left(\frac{T}{T_B} \right)^{3/2}$ both satisfy $\lim_{T \to 0} C(T, X) = 0$. Indeed, they are power laws with α=1 and α=3/2 respectively.

Even within a purely classical setting, the density of a classical ideal gas at fixed particle number becomes arbitrarily high as T goes to zero, so the interparticle spacing goes to zero. The assumption of non-interacting particles presumably breaks down when they are sufficiently close together, so the value of C_V gets modified away from its ideal constant value.

Vapor Pressure

The only liquids near absolute zero are ^{3}He and ^{4}He. Their heat of evaporation has a limiting value given by:

$$L = L_0 + C_p T$$

with L_0 and C_p constant. If we consider a container, partly filled with liquid and partly gas, the entropy of the liquid–gas mixture is:

$$S(T, x) = S_l(T) + x\left(\frac{L_0}{T} + C_p\right)$$

where S_l(T) is the entropy of the liquid and x is the gas fraction. Clearly the entropy change during the liquid–gas transition (x from 0 to 1) diverges in the limit of $T \to 0$. Nature solves this paradox as follows: at temperatures below about 50 mK the vapor pressure is so low that the gas density is lower than the best vacuum in the universe. In other words: below 50 mK there is simply no gas above the liquid.

Latent Heat of Melting

The melting curves of ^{3}He and ^{4}He both extend down to absolute zero at finite pressure. At the melting pressure, liquid and solid are in equilibrium. The third law demands that the entropies of the solid and liquid are equal at T=0. As a result, the latent heat of melting is zero and the slope of the melting curve extrapolates to zero as a result of the Clausius–Clapeyron equation.

Thermal Expansion Coefficient

The thermal expansion coefficient is defined as:

$$\alpha_V = \frac{1}{V_m}\left(\frac{\partial V_m}{\partial T}\right)_p.$$

With the Maxwell relation:

$$\left(\frac{\partial V_m}{\partial T}\right)_p = -\left(\frac{\partial S_m}{\partial p}\right)_T$$

and with X=p it is shown that:

$$\lim_{T\to 0} \alpha_V = 0.$$

So the thermal expansion coefficient of all materials must go to zero at zero kelvin.

Enthalpy

Enthalpy, a property of a thermodynamic system, is equal to the system's internal energy plus the product of its pressure and volume. In a system enclosed so as to prevent mass transfer, for processes at constant pressure, the heat absorbed or released equals the change in enthalpy.

The unit of measurement for enthalpy in the International System of Units (SI) is the joule. Other historical conventional units still in use include the British thermal unit (BTU) and the calorie.

Enthalpy comprises a system's internal energy, which is the energy required to create the system, plus the amount of work required to make room for it by displacing its environment and establishing its volume and pressure.

Enthalpy is defined as a state function that depends only on the prevailing equilibrium state identified by the system's internal energy, pressure, and volume. It is an extensive quantity.

Enthalpy(ΔH) is the preferred expression of system energy changes in many chemical, biological, and physical measurements at constant pressure, because it simplifies the description of energy transfer. In a system enclosed so as to prevent matter transfer, at constant pressure, the enthalpy change equals the energy transferred from the environment through heat transfer or work other than expansion work.

The total enthalpy, H, of a system cannot be measured directly. The same situation exists in classical mechanics: only a change or difference in energy carries physical meaning. Enthalpy itself is a thermodynamic potential, so in order to measure the enthalpy of a system, we must refer to a defined reference point; therefore what we measure is the change in enthalpy, ΔH. The ΔH is a positive change in endothermic reactions, and negative in heat-releasing exothermic processes.

For processes under constant pressure, ΔH is equal to the change in the internal energy of the system, plus the pressure-volume work $p\,\Delta V$ done by the system on its surroundings (which is > 0 for an expansion and < 0 for a contraction). This means that the change in enthalpy under such conditions is the heat absorbed or released by the system through a chemical reaction or by external heat transfer. Enthalpies for chemical substances at constant pressure usually refer to standard state: most commonly 1 bar pressure. Standard state does not, strictly speaking, specify a temperature, but expressions for enthalpy generally reference the standard heat of formation at 25 °C.

Enthalpy of ideal gases and incompressible solids and liquids does not depend on pressure, unlike entropy and Gibbs energy. Real materials at common temperatures and pressures usually closely approximate this behavior, which greatly simplifies enthalpy calculation and use in practical designs and analyses.

The word *enthalpy* was coined relatively late, in the early 20th century, in analogy with the 19th-century terms *energy* and *entropy*. The term does in fact stand in for the older term "heat content", a term which is now mostly deprecated as misleading, as dH refers to the amount of heat absorbed in a process at constant pressure only, but not in the general case (when pressure is variable). Josiah Willard Gibbs used the term "a heat function for constant pressure" for clarity.

Introduction of the concept of "heat content" H is associated with Benoît Paul Émile Clapeyron and Rudolf Clausius.

The term *enthalpy* first appeared in print in 1909. It is attributed to Heike Kamerlingh Onnes, who most likely introduced it orally the year before, at the first meeting of the Institute of Refrigeration in Paris. It gained currency only in the 1920s, notably with the *Mollier Steam Tables and Diagrams*, published in 1927.

Until the 1920s, the symbol H was used, somewhat inconsistently, for "heat" in general. The definition of H as strictly limited to enthalpy or "heat content at constant pressure" was formally proposed by Alfred W. Porter in 1922.

The enthalpy of a thermodynamic system is defined as:

$$H = U + pV,$$

where:

- H is enthalpy
- U is the internal energy of the system

- p is pressure
- V is the volume of the system

Enthalpy is an extensive property. This means that, for homogeneous systems, the enthalpy is proportional to the size of the system. It is convenient to introduce the specific enthalpy $h = \frac{H}{m}$, where m is the mass of the system, or the molar enthalpy $H_m = \frac{H}{n}$, where n is the number of moles (h and H_m are intensive properties). For inhomogeneous systems the enthalpy is the sum of the enthalpies of the composing subsystems:

$$H = \sum_k H_k,$$

where:

- H is the total enthalpy of all the subsystems
- k refers to the various subsystems
- H_k refers to the enthalpy of each subsystem
- $\sum_k H_k$ refers to the sum of the enthalpies of all subsystems

A closed system may lie in thermodynamic equilibrium in a static gravitational field, so that its pressure, p, varies continuously with altitude, while, because of the equilibrium requirement, its temperature T is invariant with altitude. (Correspondingly, the system's gravitational potential energy density also varies with altitude.) Then the enthalpy summation becomes an integral:

$$H = \int (\rho h) dV,$$

where:

- ρ ("rho") is density (mass per unit volume).
- h is the specific enthalpy (enthalpy per unit mass).
- (ρh) represents the enthalpy density (enthalpy per unit volume).
- dV, denotes an infinitesimally small element of volume within the system, for example the volume of an infinitesimally thin horizontal layer.
- $\int (\rho h) dV$ represents the sum of the enthalpies of all the elements of the volume.

The enthalpy of a closed homogeneous system is its cardinal energy function, $H(S, p)$, with natural state variables its entropy $S[p]$ and its pressure p. A differential relation for it can be derived as follows. We start from the first law of thermodynamics for closed systems for an infinitesimal process:

$$dU = \delta Q - \delta W.$$

where:

- δQ is a small amount of heat added to the system.
- δW is a small amount of work performed by the system.

In a homogeneous system in which only reversible, or quasi-static, processes are considered, the second law of thermodynamics gives $\delta Q = T\,dS$ with T the absolute temperature of the system. Furthermore, if only pV work is done, $\delta W = p\,dV$. As a result,

$$dU = T\,dS - p\,dV.$$

Adding $d(pV)$ to both sides of this expression gives:

$$dU + d(pV) = T\,dS - p\,dV + d(pV),$$

or

$$d(U + pV) = T\,dS + V\,dp.$$

So,

$$dH(S,p) = T\,dS + V\,dp.$$

Other Expressions

The above expression of dH in terms of entropy and pressure may be unfamiliar to some readers. However, there are expressions in terms of more familiar variables such as temperature and pressure:

$$dH = C_p\,dT + V(1 - \alpha T)dp.$$

Here C_p is the heat capacity at constant pressure and α is the coefficient of (cubic) thermal expansion:

$$\alpha = \frac{1}{V}\left(\frac{\partial V}{\partial T}\right)_p.$$

With this expression one can, in principle, determine the enthalpy if C_p and V are known as functions of p and T.

Note that for an ideal gas, $\alpha T = 1$, so that:

$$dH = C_p\,dT.$$

In a more general form, the first law describes the internal energy with additional terms involving the chemical potential and the number of particles of various types. The differential statement for dH then becomes:

$$dH = T\,dS + V\,dp + \sum_i \mu_i\,dN_i,$$

where μ_i is the chemical potential per particle for an *i*-type particle, and N_i is the number of such particles. The last term can also be written as $\mu_i dn_i$ (with dn_i the number of moles of component *i* added to the system and, in this case, μ_i the molar chemical potential) or as $\mu_i dm_i$ (with dm_i the mass of component *i* added to the system and, in this case, μ_i the specific chemical potential).

Cardinal Functions

The enthalpy, $H(S[p],p,\{N_i\})$, expresses the thermodynamics of a system in the *energy-language*, or in the *energy representation*. As a function of state, its arguments include both one intensive and several extensive state variables. The state variables $S[p]$, p, and $\{N_i\}$ are said to be the *natural state variables* in this representation. They are suitable for describing processes in which they are experimentally controlled. For example, in an idealized process, $S[p]$ and p can be controlled by preventing heat and matter transfer by enclosing the system with a wall that is adiathermal and impermeable to matter, and by making the process infinitely slow, and by varying only the external pressure on the piston that controls the volume of the system. This is the basis of the so-called adiabatic approximation that is used in meteorology.

Alongside the enthalpy, with these arguments, the other cardinal function of state of a thermodynamic system is its entropy, as a function, $S[p](H,p,\{N_i\})$, of the same list of variables of state, except that the entropy, $S[p]$, is replaced in the list by the enthalpy, H. It expresses the *entropy representation*. The state variables H, p, and $\{N_i\}$ are said to be the *natural state variables* in this representation. They are suitable for describing processes in which they are experimentally controlled. For example, H and p can be controlled by allowing heat transfer, and by varying only the external pressure on the piston that sets the volume of the system.

Physical Interpretation

The U term can be interpreted as the energy required to create the system, and the pV term as the work that would be required to “make room” for the system if the pressure of the environment remained constant. When a system, for example, n moles of a gas of volume V at pressure p and temperature T, is created or brought to its present state from absolute zero, energy must be supplied equal to its internal energy U plus pV, where pV is the work done in pushing against the ambient (atmospheric) pressure.

In basic physics and statistical mechanics it may be more interesting to study the internal properties of the system and therefore the internal energy is used. In basic chemistry, experiments are often conducted at constant atmospheric pressure, and the pressure-volume work represents an energy exchange with the atmosphere that cannot be accessed or controlled, so that ΔH is the expression chosen for the heat of reaction.

For a heat engine a change in its internal energy is the difference between the heat input and the pressure-volume work done by the working substance while a change in its enthalpy is the difference between the heat input and the work done by the engine:

$$dH = \delta Q - \delta W$$

where the work W done by the engine is:

$$W = -\oint p dV$$

Relationship to Heat

In order to discuss the relation between the enthalpy increase and heat supply, we return to the first law for closed systems, with the physics sign convention: $dU = \delta Q - \delta W$, where the heat δQ is supplied by conduction, radiation, and Joule heating. We apply it to the special case with a constant pressure at the surface. In this case the work term can be split into two contributions, the so-called pV work, given by $p\,dV$ (where here p is the pressure at the surface, dV is the increase of the volume of the system), and the so-called isochoric mechanical work $\delta W'$, such as stirring by a shaft with paddles or by an externally driven magnetic field acting on an internal rotor. Cases of long range electromagnetic interaction require further state variables in their formulation, and are not considered here. So we write $\delta W = p\,dV + \delta W'$. In this case the first law reads:

$$dU = \delta Q - p dV - \delta W'.$$

Now,

$$dH = dU + d(pV).$$

So,

$$dH = \delta Q + V\,dp + p dV - p dV - \delta W'$$
$$= \delta Q + V\,dp - \delta W'.$$

With sign convention of physics, $\delta W' < 0$, because isochoric shaft work done by an external device on the system adds energy to the system, and may be viewed as virtually adding heat. The only thermodynamic mechanical work done by the system is expansion work, $p\ dV$.

The system is under constant pressure ($dp = 0$). Consequently, the increase in enthalpy of the system is equal to the added heat and virtual heat:

$$dH = \delta Q - \delta W'.$$

Applications

In thermodynamics, one can calculate enthalpy by determining the requirements for creating a system from "nothingness"; the mechanical work required, pV, differs based upon the conditions that obtain during the creation of the thermodynamic system.

Energy must be supplied to remove particles from the surroundings to make space for the creation of the system, assuming that the pressure p remains constant; this is the pV term. The supplied energy must also provide the change in internal energy, U, which includes activation energies, ionization energies, mixing energies, vaporization energies, chemical bond energies, and so forth. Together, these constitute the change in the enthalpy $U + pV$. For systems at constant pressure, with no external work done other than the pV work, the change in enthalpy is the heat received by the system.

For a simple system, with a constant number of particles, the difference in enthalpy is the maximum amount of thermal energy derivable from a thermodynamic process in which the pressure is held constant.

Heat of Reaction

The total enthalpy of a system cannot be measured directly, the *enthalpy change* of a system is measured instead. Enthalpy change is defined by the following equation:

$$\Delta H = H_f - H_i,$$

where:

- ΔH is the "enthalpy change",
- H_f is the final enthalpy of the system (in a chemical reaction, the enthalpy of the products),
- H_i is the initial enthalpy of the system (in a chemical reaction, the enthalpy of the reactants).

For an exothermic reaction at constant pressure, the system's change in enthalpy equals the energy released in the reaction, including the energy retained in the system and lost through expansion against its surroundings. In a similar manner, for an endothermic reaction, the system's change in enthalpy is equal to the energy *absorbed* in the reaction, including the energy *lost by* the system and *gained* from compression from its surroundings. If ΔH is positive, the reaction is endothermic, that is heat is absorbed by the system due to the products of the reaction having a greater enthalpy than the reactants. On the other hand, if ΔH is negative, the reaction is exothermic, that is the overall decrease in enthalpy is achieved by the generation of heat.

From the definition of enthalpy as $H = U + pV$, the enthalpy change at constant pressure $\Delta H = \Delta U + p\Delta V$. However for most chemical reactions, the work term $p\Delta V$ is much smaller than the internal energy change ΔU which is approximately equal to ΔH As an example, for the combustion of carbon monoxide 2 CO(g) + O_2(g) → 2 CO_2(g), ΔH = −566.0 kJ and ΔU = −563.5 kJ. Since the differences are so small, reaction enthalpies are often loosely described as reaction energies and analyzed in terms of bond energies.

Specific Enthalpy

The specific enthalpy of a uniform system is defined as $h = H/m$ where m is the mass of the system. The SI unit for specific enthalpy is joule per kilogram. It can be expressed in other specific quantities by $h = u + pv$, where u is the specific internal energy, p is the pressure, and v is specific volume, which is equal to $1/\rho$, where ρ is the density.

Enthalpy Changes

An enthalpy change describes the change in enthalpy observed in the constituents of a thermodynamic system when undergoing a transformation or chemical reaction. It is the difference between the enthalpy after the process has completed, i.e. the enthalpy of the products, and the initial enthalpy of the system, i.e. the reactants. These processes are reversible and the enthalpy for the reverse process is the negative value of the forward change.

A common standard enthalpy change is the enthalpy of formation, which has been determined for a large number of substances. Enthalpy changes are routinely measured and compiled in chemical and physical reference works, such as the CRC Handbook of Chemistry and Physics. The following is a selection of enthalpy changes commonly recognized in thermodynamics.

When used in these recognized terms the qualifier *change* is usually dropped and the property is simply termed *enthalpy of 'process'*. Since these properties are often used as reference values it is very common to quote them for a standardized set of environmental parameters, or standard conditions, including:

- A temperature of 25 °C or 298.15 K,
- A pressure of one atmosphere (1 atm or 101.325 kPa),
- A concentration of 1.0 M when the element or compound is present in solution,
- Elements or compounds in their normal physical states, i.e. standard state.

For such standardized values the name of the enthalpy is commonly prefixed with the term *standard*, e.g. *standard enthalpy of formation*.

Chemical properties:

- Enthalpy of reaction, defined as the enthalpy change observed in a constituent of a thermodynamic system when one mole of substance reacts completely.
- Enthalpy of formation, defined as the enthalpy change observed in a constituent of a thermodynamic system when one mole of a compound is formed from its elementary antecedents.
- Enthalpy of combustion, defined as the enthalpy change observed in a constituent of a thermodynamic system when one mole of a substance burns completely with oxygen.
- Enthalpy of hydrogenation, defined as the enthalpy change observed in a constituent of a thermodynamic system when one mole of an unsaturated compound reacts completely with an excess of hydrogen to form a saturated compound.
- Enthalpy of atomization, defined as the enthalpy change required to atomize one mole of compound completely.
- Enthalpy of neutralization, defined as the enthalpy change observed in a constituent of a thermodynamic system when one mole of water is formed when an acid and a base react.
- Standard Enthalpy of solution, defined as the enthalpy change observed in a constituent of a thermodynamic system when one mole of a solute is dissolved completely in an excess of solvent, so that the solution is at infinite dilution.
- Standard enthalpy of Denaturation (biochemistry), defined as the enthalpy change required to denature one mole of compound.
- Enthalpy of hydration, defined as the enthalpy change observed when one mole of gaseous ions are completely dissolved in water forming one mole of aqueous ions.

Physical properties:

- Enthalpy of fusion, defined as the enthalpy change required to completely change the state of one mole of substance between solid and liquid states.
- Enthalpy of vaporization, defined as the enthalpy change required to completely change the state of one mole of substance between liquid and gaseous states.
- Enthalpy of sublimation, defined as the enthalpy change required to completely change the state of one mole of substance between solid and gaseous states.
- Lattice enthalpy, defined as the energy required to separate one mole of an ionic compound into separated gaseous ions to an infinite distance apart (meaning no force of attraction).
- Enthalpy of mixing, defined as the enthalpy change upon mixing of two (non-reacting) chemical substances.

Open Systems

In thermodynamic open systems, mass (of substances) may flow in and out of the system boundaries. The first law of thermodynamics for open systems states: The increase in the internal energy of a system is equal to the amount of energy added to the system by mass flowing in and by heating, minus the amount lost by mass flowing out and in the form of work done by the system:

$$dU = \delta Q + dU_{\text{in}} - dU_{\text{out}} - \delta W,$$

where U_{in} is the average internal energy entering the system, and U_{out} is the average internal energy leaving the system.

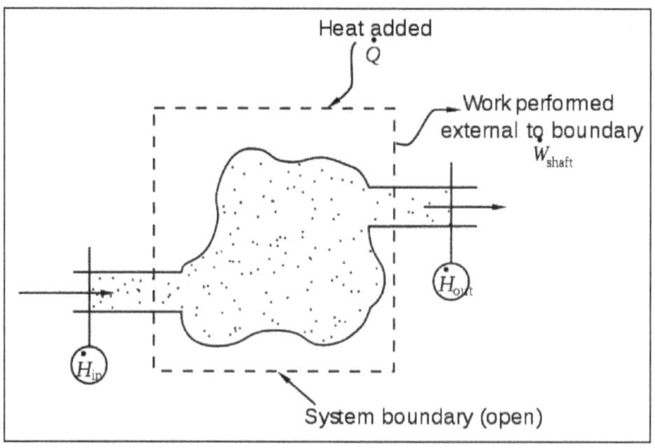

During steady, continuous operation, an energy balance applied to an open system equates shaft work performed by the system to heat added plus net enthalpy added.

The region of space enclosed by the boundaries of the open system is usually called a control volume, and it may or may not correspond to physical walls. If we choose the shape of the control volume such that all flow in or out occurs perpendicular to its surface, then the flow of mass into the system performs work as if it were a piston of fluid pushing mass into the system, and the system performs work on the flow of mass out as if it were driving a piston of fluid. There are then two types of work performed: *flow work* described above, which is performed on the fluid (this is also often called *pV work*), and *shaft work*, which may be performed on some mechanical device.

These two types of work are expressed in the equation

$$\delta W = d(p_{\text{out}}V_{\text{out}}) - d(p_{\text{in}}V_{\text{in}}) + \delta W_{\text{shaft}}.$$

Substitution into the equation above for the control volume (cv) yields:

$$dU_{\text{cv}} = \delta Q + dU_{\text{in}} + d(p_{\text{in}}V_{\text{in}}) - dU_{\text{out}} - d(p_{\text{out}}V_{\text{out}}) - \delta W_{\text{shaft}}.$$

The definition of enthalpy, H, permits us to use this thermodynamic potential to account for both internal energy and pV work in fluids for open systems:

$$dU_{\text{cv}} = \delta Q + dH_{\text{in}} - dH_{\text{out}} - \delta W_{\text{shaft}}.$$

If we allow also the system boundary to move (e.g. due to moving pistons), we get a rather general form of the first law for open systems. In terms of time derivatives it reads:

$$\frac{dU}{dt} = \sum_k \dot{Q}_k + \sum_k \dot{H}_k - \sum_k p_k \frac{dV_k}{dt} - P,$$

with sums over the various places k where heat is supplied, mass flows into the system, and boundaries are moving. The $\dot{H}_k$ terms represent enthalpy flows, which can be written as:

$$\dot{H}_k = h_k \dot{m}_k = H_{\text{m}} \dot{n}_k,$$

with $\dot{m}_k$ the mass flow and $\dot{n}_k$ the molar flow at position k respectively. The term $dV_{k/dt}$ represents the rate of change of the system volume at position k that results in pV power done by the system. The parameter P represents all other forms of power done by the system such as shaft power, but it can also be e.g. electric power produced by an electrical power plant.

Note that the previous expression holds true only if the kinetic energy flow rate is conserved between system inlet and outlet. Otherwise, it has to be included in the enthalpy balance. During steady-state operation of a device, the average dU/dt may be set equal to zero. This yields a useful expression for the average power generation for these devices in the absence of chemical reactions:

$$P = \sum_k \left\langle \dot{Q}_k \right\rangle + \sum_k \left\langle \dot{H}_k \right\rangle - \sum_k \left\langle p_k \frac{dV_k}{dt} \right\rangle,$$

where the angle brackets denote time averages. The technical importance of the enthalpy is directly related to its presence in the first law for open systems.

The enthalpy values of important substances can be obtained using commercial software. Practically all relevant material properties can be obtained either in tabular or in graphical form. There are many types of diagrams, such as h–T diagrams, which give the specific enthalpy as function of temperature for various pressures, and h–p diagrams, which give h as function of p for various T. One of the most common diagrams is the temperature–specific entropy diagram (T–s-diagram). It gives the melting curve and saturated liquid and vapor values together with isobars and isenthalps. These diagrams are powerful tools in the hands of the thermal engineer.

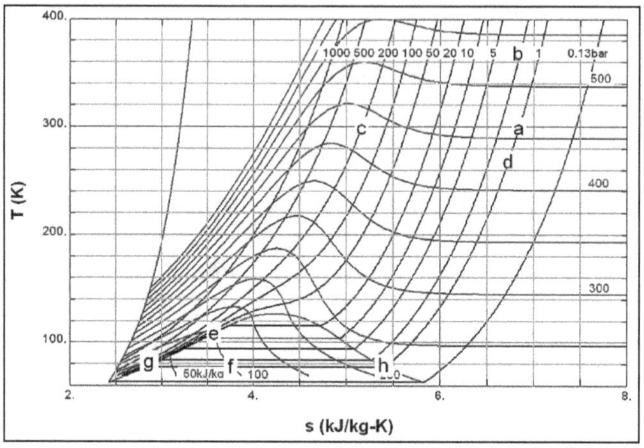

T–s diagram of nitrogen. The red curve at the left is the melting curve. The red dome represents the two-phase region with the low-entropy side the saturated liquid and the high-entropy side the saturated gas. The black curves give the *T–s* relation along isobars. The pressures are indicated in bar. The blue curves are isenthalps (curves of constant enthalpy). The values are indicated in blue in kJ/kg. The specific points a, b, etc., are treated in the main text.

Some Basic Applications

a: T = 300 K, p = 1 bar, s = 6.85 kJ/(kg K), h = 461 kJ/kg;

b: T = 380 K, p = 2 bar, s = 6.85 kJ/(kg K), h = 530 kJ/kg;

c: T = 300 K, p = 200 bar, s = 5.16 kJ/(kg K), h = 430 kJ/kg;

d: T = 270 K, p = 1 bar, s = 6.79 kJ/(kg K), h = 430 kJ/kg;

e: T = 108 K, p = 13 bar, s = 3.55 kJ/(kg K), h = 100 kJ/kg (saturated liquid at 13 bar);

f: T = 77.2 K, p = 1 bar, s = 3.75 kJ/(kg K), h = 100 kJ/kg;

g: T = 77.2 K, p = 1 bar, s = 2.83 kJ/(kg K), h = 28 kJ/kg (saturated liquid at 1 bar);

h: T = 77.2 K, p = 1 bar, s = 5.41 kJ/(kg K), h = 230 kJ/kg (saturated gas at 1 bar);

Throttling

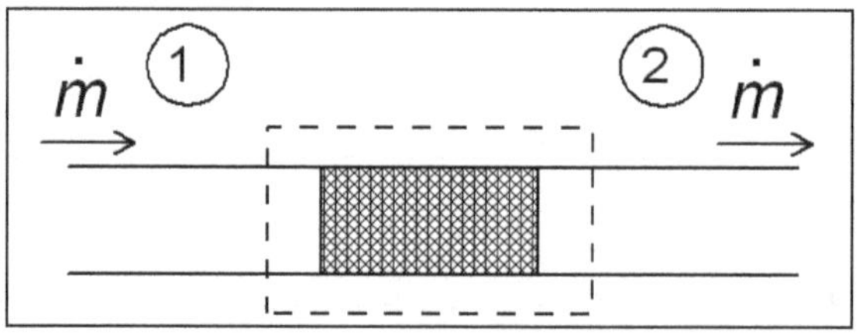

Schematic diagram of a throttling in the steady state. Fluid enters the system (dotted rectangle) at point 1 and leaves it at point 2. The mass flow is $\dot{m}$.

One of the simple applications of the concept of enthalpy is the so-called throttling process, also known as Joule-Thomson expansion. It concerns a steady adiabatic flow of a fluid through a flow resistance (valve, porous plug, or any other type of flow resistance) as shown in the figure. This process is very important, since it is at the heart of domestic refrigerators, where it is responsible for the temperature drop between ambient temperature and the interior of the refrigerator. It is also the final stage in many types of liquefiers.

For a steady state flow regime, the enthalpy of the system (dotted rectangle) has to be constant. Hence:

$$0 = \dot{m}h_1 - \dot{m}h_2.$$

Since the mass flow is constant, the specific enthalpies at the two sides of the flow resistance are the same:

$$h_1 = h_2,$$

that is, the enthalpy per unit mass does not change during the throttling. The consequences of this relation can be demonstrated using the T–s diagram above. Point c is at 200 bar and room temperature (300 K). A Joule–Thomson expansion from 200 bar to 1 bar follows a curve of constant enthalpy of roughly 425 kJ/kg lying between the 400 and 450 kJ/kg isenthalps and ends in point d, which is at a temperature of about 270 K. Hence the expansion from 200 bar to 1 bar cools nitrogen from 300 K to 270 K. In the valve, there is a lot of friction, and a lot of entropy is produced, but still the final temperature is below the starting value.

Point e is chosen so that it is on the saturated liquid line with h = 100 kJ/kg. It corresponds roughly with p = 13 bar and T = 108 K. Throttling from this point to a pressure of 1 bar ends in the two-phase region (point f). This means that a mixture of gas and liquid leaves the throttling valve. Since the enthalpy is an extensive parameter, the enthalpy in f (h_f) is equal to the enthalpy in g (h_g) multiplied by the liquid fraction in f (x_f) plus the enthalpy in h (h_h) multiplied by the gas fraction in f ($1 - x_f$). So:

$$h_f = x_f h_g + (1 - x_f)h_h.$$

With numbers: $100 = x_f \times 28 + (1 - x_f) \times 230$, so $x_f = 0.64$. This means that the mass fraction of the liquid in the liquid–gas mixture that leaves the throttling valve is 64%.

Compressors

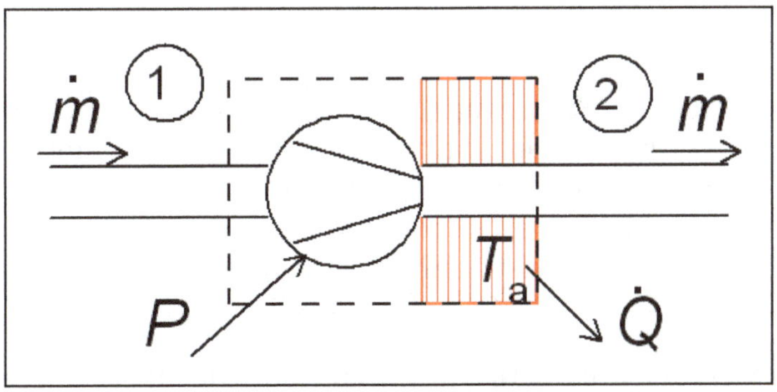

Schematic diagram of a compressor in the steady state. Fluid enters the system (dotted rectangle) at point 1 and leaves it at point 2. The mass flow is $\dot{m}$. A power P is applied and a heat flow Q is released to the surroundings at ambient temperature T_a.

A power P is applied e.g. as electrical power. If the compression is adiabatic, the gas temperature goes up. In the reversible case it would be at constant entropy, which corresponds with a vertical line in the T–s diagram. For example, compressing nitrogen from 1 bar (point a) to 2 bar (point b) would result in a temperature increase from 300 K to 380 K. In order to let the compressed gas exit at ambient temperature T_a, heat exchange, e.g. by cooling water, is necessary. In the ideal case the compression is isothermal. The average heat flow to the surroundings is Q. Since the system is in the steady state the first law gives:

$$0 = -\dot{Q} + \dot{m}h_1 - \dot{m}h_2 + P.$$

The minimal power needed for the compression is realized if the compression is reversible. In that case the second law of thermodynamics for open systems gives:

$$0 = -\frac{\dot{Q}}{T_a} + \dot{m}s_1 - \dot{m}s_2.$$

Eliminating Q gives for the minimal power:

$$\frac{P_{min}}{\dot{m}} = h_2 - h_1 - T_a(s_2 - s_1).$$

For example, compressing 1 kg of nitrogen from 1 bar to 200 bar costs at least $(h_c - h_a) - T_a(s_c - s_a)$. With the data, obtained with the T–s diagram, we find a value of (430 – 461) – 300 × (5.16 – 6.85) = 476 kJ/kg.

The relation for the power can be further simplified by writing it as:

$$\frac{P_{min}}{\dot{m}} = \int_1^2 (dh - T_a\, ds).$$

With $dh = T ds + v dp$, this results in the final relation:

$$\frac{P_{min}}{\dot{m}} = \int_1^2 v dp.$$

Entropy

Entropy is the measure of a system's thermal energy per unit temperature that is unavailable for doing useful work. Because work is obtained from ordered molecular motion, the amount of entropy is also a measure of the molecular disorder, or randomness, of a system. The concept of entropy provides deep insight into the direction of spontaneous change for many everyday phenomena. Its introduction by the German physicist Rudolf Clausius in 1850 is a highlight of 19th-century physics.

The idea of entropy provides a mathematical way to encode the intuitive notion of which processes are impossible, even though they would not violate the fundamental law of conservation of energy. For example, a block of ice placed on a hot stove surely melts, while the stove grows cooler. Such a process is called irreversible because no slight change will cause the melted water to turn back into ice while the stove grows hotter. In contrast, a block of ice placed in an ice-water bath will either thaw a little more or freeze a little more, depending on whether a small amount of heat is added to or subtracted from the system. Such a process is reversible because only an infinitesimal amount of heat is needed to change its direction from progressive freezing to progressive thawing. Similarly, compressed gas confined in a cylinder could either expand freely into the atmosphere if a valve were opened (an irreversible process), or it could do useful work by pushing a moveable piston against the force needed to confine the gas. The latter process is reversible because only a slight increase in the restraining force could reverse the direction of the process from expansion to compression. For reversible processes the system is in equilibrium with its environment, while for irreversible processes it is not.

To provide a quantitative measure for the direction of spontaneous change, Clausius introduced the concept of entropy as a precise way of expressing the second law of thermodynamics. The Clausius form of the second law states that spontaneous change for an irreversible process in an isolated system (that is, one that does not exchange heat or work with its surroundings) always proceeds in the direction of increasing entropy. For example, the block of ice and the stove constitute two parts of an isolated system for which total entropy increases as the ice melts.

By the Clausius definition, if an amount of heat Q flows into a large heat reservoir at temperature T above absolute zero, then the entropy increase is $\Delta S = Q/T$. This equation effectively gives an alternate definition of temperature that agrees with the usual definition. Assume that there are two heat reservoirs R_1 and R_2 at temperatures T_1 and T_2 (such as the stove and the block of ice). If an amount of heat Q flows from R_1 to R_2, then the net entropy change for the two reservoirs is:

$$\Delta S = Q\left(\frac{1}{T_2} - \frac{1}{T_1}\right),$$

which is positive provided that $T_1 > T_2$. Thus, the observation that heat never flows spontaneously from cold to hot is equivalent to requiring the net entropy change to be positive for a spontaneous flow of heat. If $T_1 = T_2$, then the reservoirs are in equilibrium, no heat flows, and $\Delta S = 0$.

The condition $\Delta S \geq 0$ determines the maximum possible efficiency of heat engines—that is, systems such as gasoline or steam engines that can do work in a cyclic fashion. Suppose a heat engine absorbs heat Q_1 from R_1 and exhausts heat Q_2 to R_2 for each complete cycle. By conservation of energy, the work done per cycle is $W = Q_1 - Q_2$, and the net entropy change is:

$$\Delta S = \frac{Q_2}{T_2} - \frac{Q_1}{T_1},$$

To make W as large as possible, Q_2 should be as small as possible relative to Q_1. However, Q_2 cannot be zero, because this would make ΔS negative and so violate the second law. The smallest possible value of Q_2 corresponds to the condition $\Delta S = 0$, yielding:

$$\left(\frac{Q_2}{Q_1}\right)_{\min} = \frac{T_2}{T_1}$$

as the fundamental equation limiting the efficiency of all heat engines. A process for which

$\Delta S = 0$ is reversible because an infinitesimal change would be sufficient to make the heat engine run backward as a refrigerator.

The same reasoning can also determine the entropy change for the working substance in the heat engine, such as a gas in a cylinder with a movable piston. If the gas absorbs an incremental amount of heat dQ from a heat reservoir at temperature T and expands reversibly against the maximum possible restraining pressure P, then it does the maximum work dW = PdV, where dV is the change in volume. The internal energy of the gas might also change by an amount dU as it expands. Then by conservation of energy, dQ = dU + P dV. Because the net entropy change for the system plus reservoir is zero when maximum work is done and the entropy of the reservoir decreases by an amount $dS_{reservoir} = -dQ/T$, this must be counterbalanced by an entropy increase of:

$$dS_{system} = \frac{dU + PdV}{T} = \frac{dQ}{T}$$

for the working gas so that $dS_{system} + d_{Sreservoir} = 0$. For any real process, less than the maximum work would be done (because of friction, for example), and so the actual amount of heat dQ′ absorbed from the heat reservoir would be less than the maximum amount dQ. For example, the gas could be allowed to expand freely into a vacuum and do no work at all. Therefore, it can be stated that:

$$dS_{system} = \frac{dU + PdV}{T} \geq \frac{dQ'}{T},$$

with $dQ' = dQ$ in the case of maximum work corresponding to a reversible process.

This equation defines S_{system} as a thermodynamic state variable, meaning that its value is completely determined by the current state of the system and not by how the system reached that state. Entropy is an extensive property in that its magnitude depends on the amount of material in the system.

In one statistical interpretation of entropy, it is found that for a very large system in thermodynamic equilibrium, entropy S is proportional to the natural logarithm of a quantity Ω representing the maximum number of microscopic ways in which the macroscopic state corresponding to S can be realized; that is, $S = k \ln \Omega$, in which k is the Boltzmann constant that is related to molecular energy.

All spontaneous processes are irreversible; hence, it has been said that the entropy of the universe is increasing: that is, more and more energy becomes unavailable for conversion into work. Because of this, the universe is said to be "running down."

Applications of Energy Equations in Thermodynamics

The following points highlight the major five applications of energy equations in thermodynamics. The applications are: 1. Boiler 2. Turbine (Engine)/Compressor (or Pump) 3. Throttling Process 4. Nozzle 5. Condenser.

Boiler

It is a steam generator.

Characteristics of the Boiler

- Change in velocity of the fluid at the entrance and exit is very small, and hence ΔKE can be neglected, i.e., ΔKE = 0.
- Change in elevation between the inlet and exit point is very small so change in PE may be neglected, i.e., ΔPE = 0.
- Since no work is done in a boiler w = 0.

∴ From SFEE on mass basis,

$$\left[\frac{C_1^2}{2}+gZ_1+h_1\right]+q=\left[\frac{C_2^2}{2}+gZ_2+h_2\right]+w$$

i.e., $$q=\left[\frac{C_2^2-C_1^2}{2}+g(Z_2+Z_1)+(h_2+h_1)\right]+w$$

On substituting the above condition in equation become,

$$q=\left[0+0+(h_2+h_1)\right]+0$$

i.e., $$q = h_2 - h_{1ss}$$

Turbine (Engine)/Compressor (or Pump)

The turbines and engines are power producing devices whereas Compressors, Pumps, Blowers etc. are power consuming devices.

Characteristic of a Turbine

- Since it is insulated, negligible heat transfer.

  ∴ q = 0

- Change in velocity of the fluid at the entrance and at the exit is negligible.

  ∴ ΔKE = 0

- Since change in elevation between inlet and exit point is negligible,

  ∴ From SFEE:

i.e., $$q=\left[\frac{(C_2^2-C_1^2)}{2}+g(Z_2-Z_1)+(h_2-h_1)\right]+w$$

∴ The equation reduces to, $0=\left[0+0+(h_2-h_1)\right]+w$

or $$w = h_1 - h_2$$

Here work is produced at the expense of enthalpy. Similarly for an Adiabatic Pump/Compressor, work is done on the system, so w is -ve.

∴ SFEE becomes, $w = h_2 - h_1$ (since w is –ve).

Throttling Process

When a fluid flows through a partially opened valve or an orifice then there is an appreciable pressure drop and the flow is said to be throttled.

Characteristics

- $q = 0$,
- $w = 0$,
- $\Delta PE = 0$.

Substituting these in SFEE we get:

$$q = \left[\frac{(C_2^2 - C_1^2)}{2} + g(Z_2 - Z_1) + (h_2 - h_1)\right] + w$$

$$0 = \left[\frac{(C_2^2 - C_1^2)}{2} + 0 + (h_2 - h_1)\right] + 0$$

$$\text{i.e.,} \quad \frac{C_1^2}{2} + h_1 = \frac{C_2^2}{2} + h_2$$

Nozzle

It is a device which is used to increase the KE of flowing fluids:

- $q = 0$
- $w = 0$
- $\Delta PE = 0$

∴ Substituting these in SFEE we get,

$$q = \left[\frac{(C_2^2 - C_1^2)}{2} + g(Z_2 - Z_1) + (h_2 - h_1)\right] + w$$

$$0 = \left[\frac{(C_2^2 - C_1^2)}{2} + 0 + (h_2 - h_1)\right] + 0$$

$$\text{i.e.,} \quad \frac{C_1^2}{2} + h_1 = \frac{C_2^2}{2} + h_2$$

Condenser

Condenser or to show in a steady flow process with multiple streams of fluid without mixing.

$$\frac{\dot{m}_1}{\dot{m}_3} = \frac{h_4 - h_3}{h_1 - h_2}$$

In this case, cooling water enters at section 1–1' passes through the tubes and leaves the condenser at section 2–2'.

Steam to be condensed enters the condenser at Section 3-3 and when it comes in contact with the cold surface of water tubes, it gets condensed and the condensate leaves the condenser at Section 4-4'.

So, in this case working fluids are water and steam i.e., two fluids flow through the system and since they do not mix with each other, it may be considered as a steady flow system with multiple streams of fluid without mixing.

Let, $\dot{m}_1\ \dot{m}_2$ – Mass flow rates of cooling water at Secs.1-1' and 2-2' in kg/sec.

$\dot{m}_3\ \dot{m}_4$ – Mass flow rates of at steam and condenstate Secs.3-3' and 4-4' in kg/sec.

$h_1\ h_2$ – Specific enthalpics of cooling water at Secs.1-1' and 2-2' respectively in j/kg.

$h_3\ h_4$ – Specific enthalpics of steam and condenstate at Secs.3-3' and 4-4' respectively in J/kg.

Since it is steady flow system, writing SFEE for the coolant.

$$\dot{m}_1\left[\frac{C_1^2}{2} + gZ_1 + h_1\right] + \frac{\delta Q}{dt} = \dot{m}_2\left[\frac{C_2^2}{2} + gZ_2 + h_2\right] + \frac{\delta w}{dt}$$

And for steam,

$$\dot{m}_3\left[\frac{C_3^2}{2} + gZ_3 + h_3\right] + \frac{\delta Q}{dt} = \dot{m}_4\left[\frac{C_4^2}{2} + gZ_4 + h_4\right] + \frac{\delta w}{dt}$$

Neglecting the change in KE and PE, since they are very small and as the condenser is insulated δQ/dt=0 and as no work is done in the condenser δW/t=0.

So, equations above reduce to,

$$\dot{m}_1\ h_1 = \dot{m}_2\ h_2$$
$$\dot{m}_3\ h_3 = \dot{m}_4\ h_4$$

Now since it is a steady flow system:

$$\begin{bmatrix}\text{Total energy entering the}\\ \text{condenser with steam}\\ \text{and coolant}\end{bmatrix} = \begin{bmatrix}\text{Total energy leaving the}\\ \text{condenser with coolant}\\ \text{and condensate}\end{bmatrix}$$

$$\therefore \dot{m}_1 h_1 + \dot{m}_3 h_3 = \dot{m}_2 h_2 + \dot{m}_4 h_4$$

But for a steady flow system:

$$\dot{m}_1 = \dot{m}_2 \ and \ \dot{m}_3 = \dot{m}_4$$

$$\therefore \dot{m}_1 (h_1 - h_2) = \dot{m}_3 (h_4 - h_3)$$

$$\therefore \frac{\dot{m}_1}{\dot{m}_3} = \frac{h_4 - h_3}{h_1 - h_2}. \text{ Hence proved.}$$

References

- What-is-thermal-engineering, topic: whatisengineering.com, Retrieved 18 June, 2019
- Gislason, E. A.; Craig, N. C. (2005). "Cementing the foundations of thermodynamics:comparison of system-based and surroundings-based definitions of work and heat". J. Chem. Thermodynamics. 37 (9): 954–966. doi:10.1016/j.jct.2004.12.012
- Heat, definition: whatis.techtarget.com, Retrieved 19 July, 2019
- Pidwirny, Michael (2008). "CHAPTER 8: Introduction to the Hydrosphere (e). Cloud Formation Processes". Physical Geography. Archived from the original on 2008-12-20. Retrieved 2009-01-01
- Thermodynamics, science: britannica.com, Retrieved 20 August, 2019
- Sam Zhang; Dongliang Zhao (19 November 2012). Aeronautical and Aerospace Materials Handbook. CRC Press. Pp. 304–. ISBN 978-1-4398-7329-8. Retrieved 7
- Entropy-physics, science: britannica.com, Retrieved 21 January, 2019
- Pokrovskii V.N. (2013) A derivation of the main relations of non-equilibrium thermodynamics. Hinda-wi Publishing Corporation: ISRN Thermodynamics, vol. 2013, article ID 906136, 9 p. Https://dx.doi.org/10.1155/2013/906136
- Top-5-applications-of-energy-equations-in-thermodynamics-thermodynamics, thermodynamics, thermal-engineering: engineeringenotes.com, Retrieved 22 February, 2019
- Weisstein, Eric W. "Radiation". Eric Weisstein's World of Physics. Wolfram Research. Retrieved 11 January 2014

2
Steam Power Cycles

Many different types of cycles are used to operate a steam power cycle. Some of these cycles are Rankine cycle, Carnot cycle, binary cycle, Diesel cycle, Atkinson cycle, Brayton cycle, etc. The topics elaborated in this chapter will help in gaining a better perspective about the various operations of these steam power cycles.

Carnot Cycle

The Carnot cycle consists of the following four processes:

1. A reversible isothermal gas expansion process. In this process, the ideal gas in the system absorbs q_{in} amount heat from a heat source at a high temperature T_h, expands and does work on surroundings.

2. A reversible adiabatic gas expansion process. In this process, the system is thermally insulated. The gas continues to expand and do work on surroundings, which causes the system to cool to a lower temperature, T_l.

3. A reversible isothermal gas compression process. In this process, surroundings do work to the gas at T_l, and causes a loss of heat, q_{out}.

4. A reversible adiabatic gas compression process. In this process, the system is thermally insulated. Surroundings continue to do work to the gas, which causes the temperature to rise back to T_h.

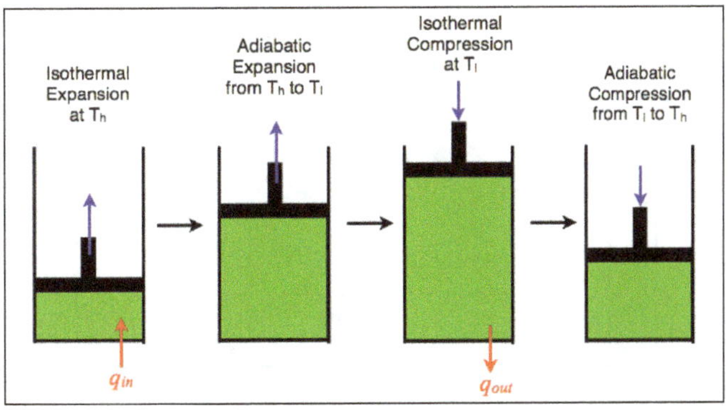

An ideal gas-piston model of the Carnot cycle.

P-V Diagram

The P-V diagram of the Carnot cycle is shown in figure. In isothermal processes I and III, ΔU=0 because ΔT=0. In adiabatic processes II and IV, q=0. Work, heat, ΔU, and ΔH of each process in the Carnot cycle is shown in the table below.

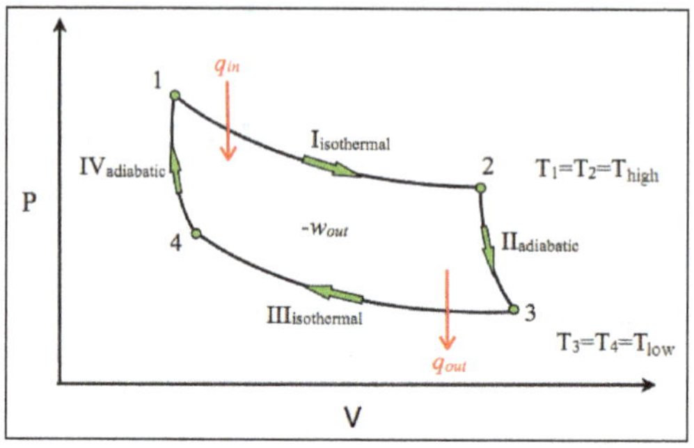

A P-V diagram of the Carnot Cycle.

Table: Work, heat, ΔU, and ΔH in the P-V diagram of the Carnot Cycle.

Process	w	q	ΔU	ΔH
I	$-nRT_{high}\ln\left(\frac{V_2}{V_1}\right)$	$nRT_{high}\ln\left(\frac{V_2}{V_1}\right)$	0	0
II	$n\bar{C}_v(T_{low}-T_{high})$	0	$n\bar{C}_v(T_{low}-T_{high})$	$n\bar{C}_p(T_{low}-T_{high})$
III	$-nRT_{low}\ln\left(\frac{V_4}{V_3}\right)$	$nRT_{low}\ln\left(\frac{V_4}{V_3}\right)$	0	0
IV	$n\bar{C}_v(T_{hight}-T_{low})$	0	$n\bar{C}_v(T_{hight}-T_{low})$	$n\bar{C}_p(T_{hight}-T_{low})$
Full Cycle	$-nRT_{high}\ln\left(\frac{V_2}{V_1}\right)-nRT_{low}\ln\left(\frac{V_4}{V_3}\right)$	$nRT_{high}\ln\left(\frac{V_2}{V_1}\right)-nRT_{low}\ln\left(\frac{V_4}{V_3}\right)$	0	0

T-S Diagram

The T-S diagram of the Carnot cycle is shown in figure. In isothermal processes I and III, ΔT=0. In adiabatic processes II and IV, ΔS=0 because dq=0.

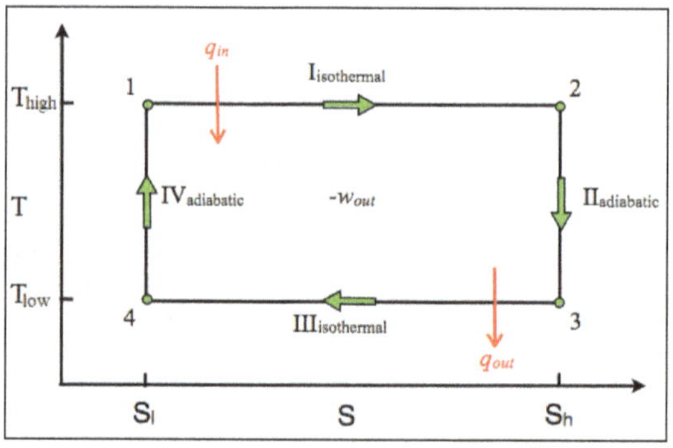

A T-S diagram of the Carnot Cycle.

ΔT and ΔS of each process in the Carnot cycle are shown in table.

Table: Work, heat, and ΔU in the T-S diagram of the Carnot Cycle.

Process	ΔT	ΔS
I	0	$nR\ln\left(\frac{}{}\right)$
II	$T_{low} - T_{high}$	0
III	0	$-nR\ln\left(\frac{V_4}{V_3}\right)$
IV	$T_{high} - T_{low}$	0
Full Cycle	0	0

Efficiency

The Carnot cycle is the most efficient engine possible based on the assumption of the absence of incidental wasteful processes such as friction, and the assumption of no conduction of heat between different parts of the engine at different temperatures. The efficiency of the carnot engine is defined as the ratio of the energy output to the energy input.

$$\text{efficiency} = \frac{\text{net work done by heat engine}}{\text{heat absorbed by heat engine}} = \frac{-\omega_{sys}}{q_{high}}$$

$$= \frac{nRT_{high}\ln\left(\frac{V_2}{V_1}\right) + nRT_{low}\ln\left(\frac{V_4}{V_3}\right)}{nRT_{high}\ln\left(\frac{V_2}{V_1}\right)}$$

Since processes II and IV are adiabatic,

$$\left(\frac{T_2}{T_3}\right)^{Cv/R} = \frac{V_3}{V_2}$$

and

$$\left(\frac{T_1}{T_4}\right)^{Cv/R} = \frac{V_4}{V_1}$$

and since $T_1 = T_2$ and $T_3 = T_{4r}$

$$\frac{V_3}{V_4} = \frac{V_2}{V_1}$$

Therefore,

$$\text{efficiency} = \frac{nRT_{high}\ln\left(\frac{V_2}{V_1}\right) - nRT_{low}\ln\left(\frac{V_2}{V_1}\right)}{nRT_{high}\ln\left(\frac{V_2}{V_1}\right)}$$

$$\text{efficiency} = \frac{T_{high} - T_{low}}{T_{high}}$$

Rankine Cycle

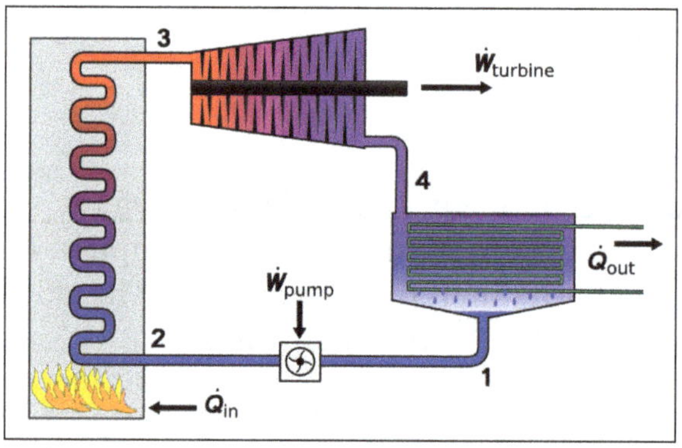

Physical layout of the Rankine cycle 1. Pump, 2. Boiler, 3. Turbine, 4. Condenser.

The Rankine cycle is a model used to predict the performance of steam turbine systems. It was also used to study the performance of reciprocating steam engines. The Rankine cycle is an idealized thermodynamic cycle of a heat engine that converts heat into mechanical work while undergoing phase change. It is an idealized cycle in which friction losses in each of the four components are neglected. The heat is supplied externally to a closed loop, which usually uses water as the working fluid. It is named after William John Macquorn Rankine, a Scottish polymath and Glasgow University professor.

The Rankine cycle closely describes the process by which steam-operated heat engines commonly found in thermal power generation plants generate power.

Power depends on the temperature difference between a heat source and a cold source. The higher the difference, the more mechanical power can be efficiently extracted out of heat energy, as per Carnot's theorem.

The heat sources used in these power plants are usually nuclear fission or the combustion of fossil fuels such as coal, natural gas, and oil, or concentrated solar power. The higher the temperature, the better.

The efficiency of the Rankine cycle is limited by the high heat of vaporization of the working fluid. Also, unless the pressure and temperature reach super critical levels in the steam boiler, the temperature range the cycle can operate over is quite small: steam turbine entry temperatures are typically around 565 °C and steam condenser temperatures are around 30 °C. This gives a theoretical maximum Carnot efficiency for the steam turbine alone of about 63.8% compared with an actual overall thermal efficiency of up to 42% for a modern coal-fired power station. This low steam turbine entry temperature (compared to a gas turbine) is why the Rankine (steam) cycle is often used as a bottoming cycle to recover otherwise rejected heat in combined-cycle gas turbine power stations.

The cold source (the colder the better) used in these power plants are usually cooling towers and a large water body (river or sea). The efficiency of the Rankine cycle is limited on the cold side by the lower practical temperature of the working fluid.

The working fluid in a Rankine cycle follows a closed loop and is reused constantly. The water vapor with condensed droplets often seen billowing from power stations is created by the cooling systems (not directly from the closed-loop Rankine power cycle). This 'exhaust' heat is represented by the "Q_{out}" flowing out of the lower side of the cycle shown in the T–s diagram below. Cooling towers operate as large heat exchangers by absorbing the latent heat of vaporization of the working fluid and simultaneously evaporating cooling water to the atmosphere.

While many substances could be used as the working fluid in the Rankine cycle, water is usually the fluid of choice due to its favorable properties, such as its non-toxic and unreactive chemistry, abundance, and low cost, as well as its thermodynamic properties. By condensing the working steam vapor to a liquid the pressure at the turbine outlet is lowered and the energy required by the feed pump consumes only 1% to 3% of the turbine output power and these factors contribute to a higher efficiency for the cycle. The benefit of this is offset by the low temperatures of steam admitted to the turbine(s). Gas turbines, for instance, have turbine entry temperatures approaching 1500 °C. However, the thermal efficiency of actual large steam power stations and large modern gas turbine stations are similar.

The Four Processes in the Rankine Cycle

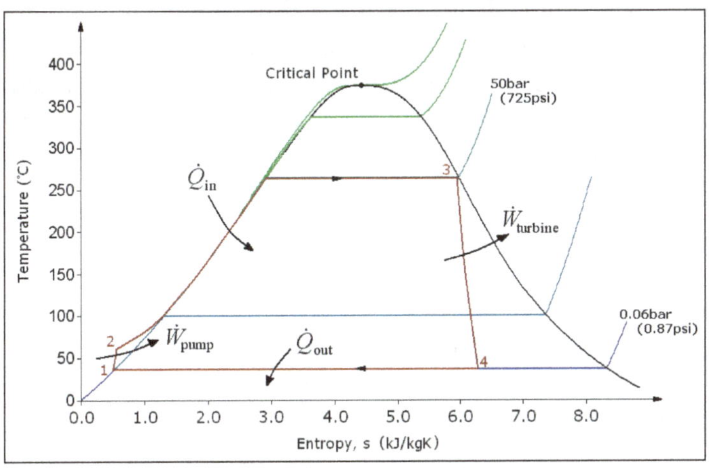

T–s diagram of a typical Rankine cycle operating between pressures of 0.06 bar and 50 bar. Left from the bell-shaped curve is liquid, right from it is gas, and under it is saturated liquid–vapour equilibrium.

There are four processes in the Rankine cycle. The states are identified by numbers (in brown) in the T–s diagram.

- Process 1–2: The working fluid is pumped from low to high pressure. As the fluid is a liquid at this stage, the pump requires little input energy.

In other words Process 1-2 is (Isentropic compression in pump).

- Process 2–3: The high-pressure liquid enters a boiler, where it is heated at constant pressure by an external heat source to become a dry saturated vapour. The input energy required can be easily calculated graphically, using an enthalpy–entropy chart (h–s chart, or Mollier diagram), or numerically, using steam tables.

In other words Process 2-3 is (Constant pressure heat addition in boiler).

- Process 3–4: The dry saturated vapour expands through a turbine, generating power. This decreases the temperature and pressure of the vapour, and some condensation may occur. The output in this process can be easily calculated using the chart.

In other words Process 3-4 is (Isentropic expansion in turbine).

- Process 4–1: The wet vapour then enters a condenser, where it is condensed at a constant pressure to become a saturated liquid.

In other words Process 4-1 is (Constant pressure heat rejection in condenser).

In an ideal Rankine cycle the pump and turbine would be isentropic, i.e., the pump and turbine would generate no entropy and hence maximize the net work output. Processes 1–2 and 3–4 would be represented by vertical lines on the T–s diagram and more closely resemble that of the Carnot cycle. The Rankine cycle shown here prevents the state of the working fluid from ending up in the superheated vapor region after the expansion in the turbine, which reduces the energy removed by the condensers.

The actual vapor power cycle differs from the ideal Rankine cycle because of irreversibilities in the inherent components caused by fluid friction and heat loss to the surroundings; fluid friction causes pressure drops in the boiler, the condenser, and the piping between the components, and as a result the steam leaves the boiler at a lower pressure; heat loss reduces the net work output, thus heat addition to the steam in the boiler is required to maintain the same level of net work output.

Variables

$\dot{Q}$	Heat flow rate to or from the system (energy per unit time).
$\dot{m}$	Mass flow rate (mass per unit time).
$\dot{W}$	Mechanical power consumed by or provided to the system (energy per unit time).
η_{therm}	Thermodynamic efficiency of the process (net power output per heat input, dimensionless).

η_{pump}, η_{turb}	Isentropic efficiency of the compression (feed pump) and expansion (turbine) processes, dimensionless.
h_1, h_2, h_3, h_4	The "specific enthalpies" at indicated points on the T–s diagram.
h_{4s}	The final "specific enthalpy" of the fluid if the turbine were isentropic.
p_1, p_2	The pressures before and after the compression process.

In general, the efficiency of a simple rankine cycle can be written as:

$$\eta_{therm} = \frac{\dot{W}_{thermal} - \dot{W}}{\dot{Q}_{in}} \approx \frac{\dot{W}_{turb}}{\dot{Q}_{in}}.$$

Each of the next four equations is derived from the energy and mass balance for a control volume. η_{therm} defines the thermodynamic efficiency of the cycle as the ratio of net power output to heat input. As the work required by the pump is often around 1% of the turbine work output, it can be simplified.

$$\frac{\dot{Q}_{in}}{\dot{m}} = h_3 - h_2,$$

$$\frac{\dot{Q}_{out}}{\dot{m}} = h_4 - h_1,$$

$$\frac{\dot{W}_{pump}}{\dot{m}} = h_2 - h_1,$$

$$\frac{\dot{W}_{turbine}}{\dot{m}} = h_3 - h_4.$$

When dealing with the efficiencies of the turbines and pumps, an adjustment to the work terms must be made:

$$\frac{\dot{W}_{pump}}{\dot{m}} = h_2 - h_1 \approx \frac{v_1 \Delta p}{\eta_{pump}} = \frac{v_1 (p_2 - p_1)}{\eta_{pump}},$$

$$\frac{\dot{W}_{turbine}}{\dot{m}} = h_3 - h_4 \approx (h_3 - h_4)\eta_{turbine}.$$

Real Rankine Cycle (non-ideal)

In a real power-plant cycle (the name "Rankine" cycle is used only for the ideal cycle), the compression by the pump and the expansion in the turbine are not isentropic. In other words, these processes are non-reversible, and entropy is increased during the two processes. This somewhat increases the power required by the pump and decreases the power generated by the turbine.

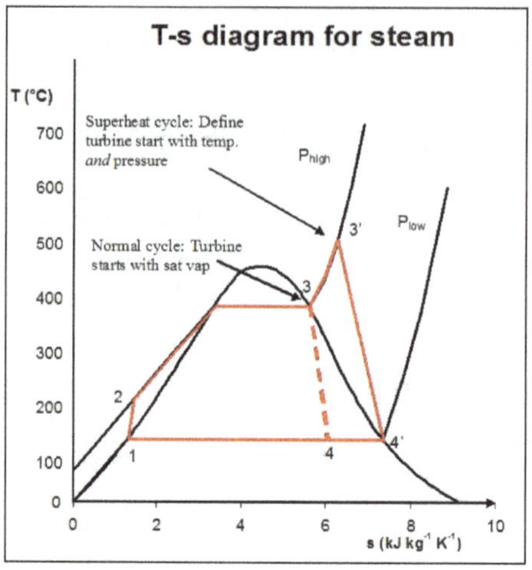

Rankine cycle with superheat.

In particular, the efficiency of the steam turbine will be limited by water-droplet formation. As the water condenses, water droplets hit the turbine blades at high speed, causing pitting and erosion, gradually decreasing the life of turbine blades and efficiency of the turbine. The easiest way to overcome this problem is by superheating the steam. On the T–s diagram above, state 3 is at a border of the two-phase region of steam and water, so after expansion the steam will be very wet. By superheating, state 3 will move to the right (and up) in the diagram and hence produce a drier steam after expansion.

Variations of the Basic Rankine Cycle

The overall thermodynamic efficiency can be increased by raising the average heat input temperature:

$$\bar{T}_{in} = \frac{\int_2^3 T\,dQ}{Q_{in}}$$

of that cycle. Increasing the temperature of the steam into the superheat region is a simple way of doing this. There are also variations of the basic Rankine cycle designed to raise the thermal efficiency of the cycle in this way.

Rankine Cycle with Reheat

The purpose of a reheating cycle is to remove the moisture carried by the steam at the final stages of the expansion process. In this variation, two turbines work in series. The first accepts vapor from the boiler at high pressure. After the vapor has passed through the first turbine, it re-enters the boiler and is reheated before passing through a second, lower-pressure, turbine. The reheat temperatures are very close or equal to the inlet temperatures, whereas the optimal reheat pressure needed is only one fourth of the original boiler pressure. Among other advantages, this prevents the vapor from condensing during its expansion and thereby reducing the damage in the turbine blades, and improves the efficiency of the cycle, because more of the heat flow into the cycle occurs at higher temperature. The reheat cycle was first introduced in the 1920s, but was not operational for long due to technical difficulties. In the 1940s, it was reintroduced with the increasing manufacture of high-pressure boilers, and eventually double reheating was introduced in the 1950s. The idea behind double reheating is to increase the average temperature. It was observed that more than two stages of reheating are unnecessary, since the next stage increases the cycle efficiency only half as much as the preceding stage. Today, double reheating is commonly used in power plants that operate under supercritical pressure.

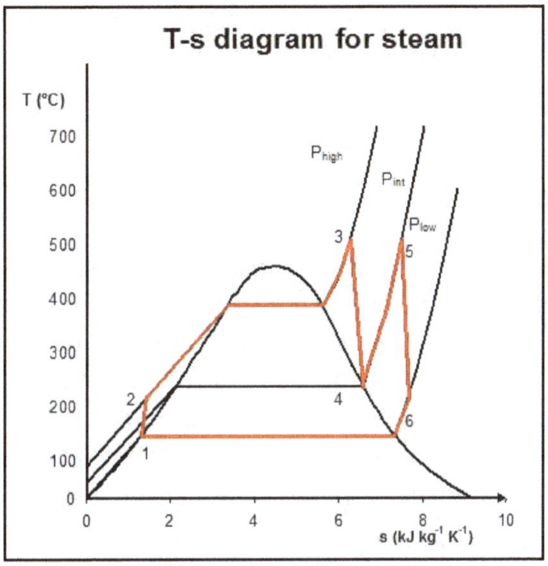

Rankine cycle with reheat.

Regenerative Rankine Cycle

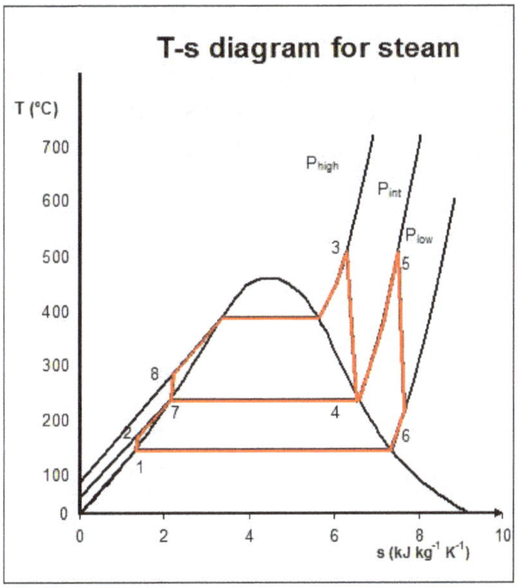

Regenerative Rankine cycle.

The regenerative Rankine cycle is so named because after emerging from the condenser (possibly as a subcooled liquid) the working fluid is heated by steam tapped from the hot portion of the cycle. In the diagram shown, the fluid at 2 is mixed with the fluid at 4 (both at the same pressure) to end up with the saturated liquid at 7. This is called "direct-contact heating". The Regenerative Rankine cycle (with minor variants) is commonly used in real power stations.

Another variation sends *bleed steam* from between turbine stages to feedwater heaters to preheat the water on its way from the condenser to the boiler. These heaters do not mix the input steam and condensate, function as an ordinary tubular heat exchanger, and are named "closed feedwater heaters".

Regeneration increases the cycle heat input temperature by eliminating the addition of heat from the boiler/fuel source at the relatively low feedwater temperatures that would exist without regenerative feedwater heating. This improves the efficiency of the cycle, as more of the heat flow into the cycle occurs at higher temperature.

Organic Rankine Cycle

The organic Rankine cycle (ORC) uses an organic fluid such as n-pentane or toluene in place of water and steam. This allows use of lower-temperature heat sources, such as solar ponds, which typically operate at around 70–90 °C. The efficiency of the cycle is much lower as a result of the lower temperature range, but this can be worthwhile because of the lower cost involved in gathering heat at this lower temperature. Alternatively, fluids can be used that have boiling points above water, and this may have thermodynamic benefits. The properties of the actual working fluid has great influence on the quality of steam (vapour) after the expansion step, influencing the design of the whole cycle.

The Rankine cycle does not restrict the working fluid in its definition, so the name "organic cycle" is simply a marketing concept and the cycle should not be regarded as a separate thermodynamic cycle.

Supercritical Rankine Cycle

The Rankine cycle applied using a supercritical fluid combines the concepts of heat regeneration and supercritical Rankine cycle into a unified process called the regenerative supercritical cycle (RGSC). It is optimised for temperature sources 125–450 °C.

Modified Rankine Cycle

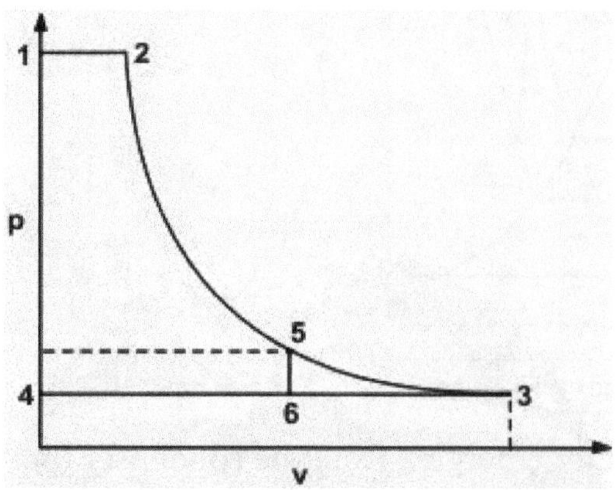

p-v diagram of modified Rankine cycle.

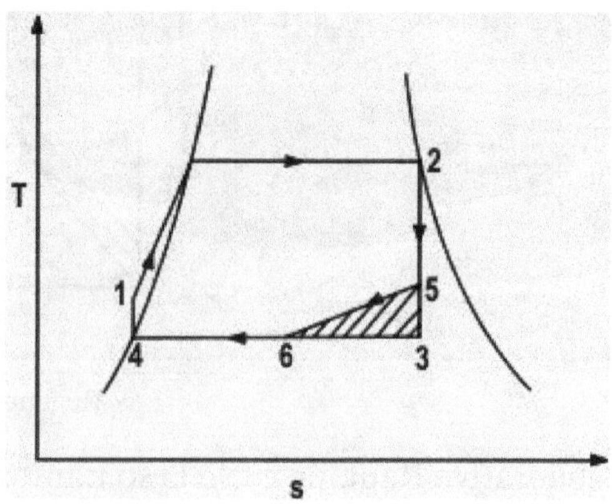

T-s diagram of modified Rankine cycle.

Process 1-2 represents the admission of high pressure steam into the engine cylinder, process 2-3 is the reversible adiabatic expansion of steam in the cylinder and process 3-4 is the exhaust of steam into condenser. Net work done is represented by the area 1- 2-3-4-1.

Observe that the area 3-6-5 is very small and in order to obtain this small work, the cylinder volume must be increased from v6 to v3.This makes cylinder very bulky. For this reason, the expansion process is terminated at point 5. So that indicator diagram becomes 1-2-5-6-4. The work lost is small but there is large saving in cylinder volume. Process 5-6 represents the release of steam into the condenser, thus causing the cylinder pressure to drop from P5 to P6. Process 6-4 is the exhaust of steam at constant pressure. Cycle 1-2-5-6-4 is called as the "modified Rankine cycle".

Thermal Efficiency:

Considering the unit mass of working fluid,

$$\text{Heat supplied} = h_2 - h_1$$

$$\begin{aligned}\text{Net workdone} &= \{W_{2\text{-}5} + W_{5\text{-}6} + W_{4-1}\}\\ &= (h_2 - h_5) - \int_5^6 vdp + (h_4 - h_1)\\ &= (h_2 - h_5) + v_5(p_5 - p_6) + (h_4 - h_1)\end{aligned}$$

V_5 = specific volume of steam at state 5.

$$\eta_{th} = \frac{\text{Net workdone}}{\text{Heat sup plied}} = \frac{(h_2 - h_5) + v_5(p_5 - p_6) + (h_4 - h_1)}{h_2 - h_1}$$

If pump work is neglected, then $h_4 \approx h_1$ ~

$$\eta_{th} = \frac{(h_2 - h_5) + v_5(p_5 - p_6)}{h_2 - h_4}$$

Binary Cycle

A binary cycle power plant is a type of geothermal power plant that allows cooler geothermal reservoirs to be used than is necessary for dry steam and flash steam plants. As of 2010, flash steam plants are the most common type of geothermal power generation plants in operation today, which use water at temperatures greater than 182 °C (455 K; 360 °F) that is pumped under high pressure to the generation equipment at the surface. With binary cycle geothermal power plants, pumps are used to pump hot water from a geothermal well, through a heat exchanger, and the cooled water is returned to the underground reservoir. A second "working" or "binary" fluid with a low boiling point, typically a butane or pentane hydrocarbon, is pumped at fairly high pressure (500 psi (3.4 MPa)) through the heat exchanger, where it is vaporized and then directed through a turbine. The vapor exiting the turbine is then condensed by cold air radiators or cold water and cycled back through the heat exchanger.

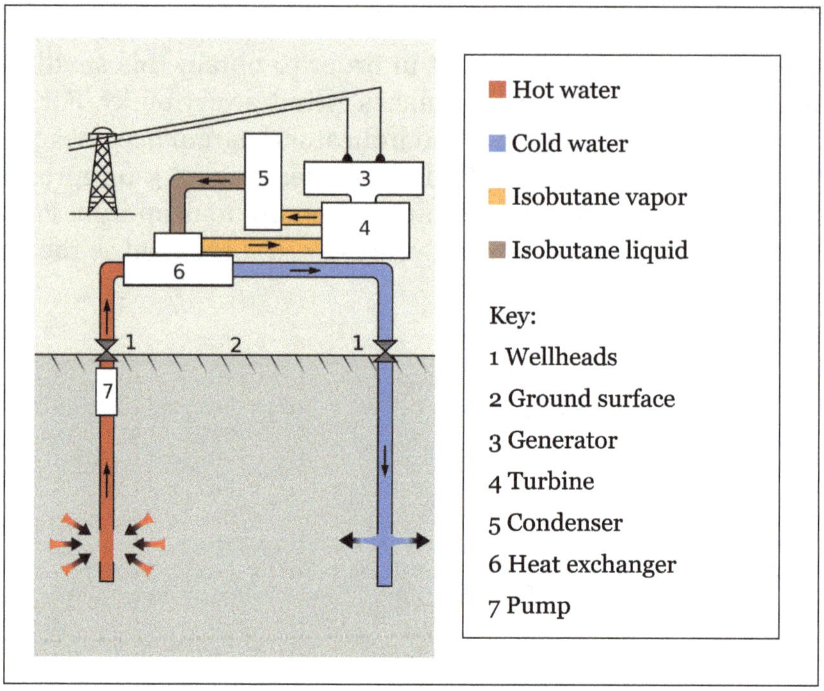

Electricity generation in a vapor-dominated hydrothermal system.

A binary vapor cycle is defined in thermodynamics as a power cycle that is a combination of two cycles, one in a high temperature region and the other in a lower temperature region.

The use of mercury-water cycles in the United States can be dated back to the late 1920s. A small mercury-water plant which produced about 40 megawatts (MW) was in use in New Hampshire in the 1950s, with a higher thermal efficiency than most of the power plants in use during the 1950s. Unfortunately, binary vapor cycles have a high initial cost and so they are not as economically attractive.

Water is the optimal working fluid to use in vapor cycles because it is the closest to an ideal working fluid that is currently available. The binary cycle is a process designed to overcome the imperfections of water as a working fluid. The cycle uses two fluids in an attempt to approach an ideal working fluid.

Characteristics of Optimal Working Fluids

Selecting the optimal working fluid has pivotal importance, since they make a significant impact on the performance of binary cycles.

1. A high critical temperature and maximum pressure.

2. Low triple-point temperature.

3. A condenser pressure that is not too low (a substance with a saturation pressure at the ambient temperature is too low).

4. A high enthalpy of vaporization (hfg).

5. A saturation dome that resembles an inverted U.

6. High thermal conductivity (good heat transfer characteristics).

7. Other properties: nontoxic, inert, inexpensive, and readily available.

Systems

Rankine Vapor Cycle

The Rankine cycle is the ideal form of a vapor power cycle. The ideal conditions can be reached by superheating the steam in the boiler and condensing it completely in the condenser. The ideal Rankine cycle does not involve any internal irreversibilities and consists of four processes; isentropic compression in a pump, constant pressure heat addition in a boiler, isentropic expansion in a turbine, and constant pressure heat rejection in a condenser.

Dual Pressure

This process is designed to reduce the thermodynamic losses incurred in the brine heat exchangers of the basic cycle. The losses occur through the process of transferring heat across a large temperature difference between the high temperature brine and the lower temperature of the working fluid. Losses are reduced by maintaining a closer match between the brine cooling curve and the working fluid heating curve.

Dual Fluid

"Power is extracted from a stream of hot fluid, such as geothermal water, by passing the stream in heat exchange relationship with a working fluid to vaporize the latter, expanding the vapor through a turbine, and condensing the vapor in a conventional Rankine cycle. Additional power is obtained in a second Rankine cycle by employing a portion of the hot fluid after heat exchange with the working fluid to vaporize a second working fluid having a lower boiling point and higher vapor density than the first fluid."

Otto Cycle

An Otto cycle is an idealized thermodynamic cycle that describes the functioning of a typical spark ignition piston engine. It is the thermodynamic cycle most commonly found in automobile engines.

The Otto cycle is a description of what happens to a mass of gas as it is subjected to changes of pressure, temperature, volume, addition of heat, and removal of heat. The mass of gas that is subjected to those changes is called the system. The system, in this case, is defined to be the fluid (gas) within the cylinder. By describing the changes that take place within the system, it will also describe in inverse, the system's effect on the environment. In the case of the Otto cycle, the effect will be to produce enough net work from the system so as to propel an automobile and its occupants in the environment.

The Otto cycle is constructed from:

- Top and bottom of the loop: a pair of quasi-parallel and isentropic processes (frictionless, adiabatic reversible).
- Left and right sides of the loop: a pair of parallel isochoric processes (constant volume).

The isentropic process of compression or expansion implies that there will be no inefficiency (loss of mechanical energy), and there be no transfer of heat into or out of the system during that process. The cylinder and piston are assumed to be impermeable to heat during that time. Work is performed on the system during the lower isentropic compression process. Heat flows into the Otto cycle through the left pressurizing process and some of it flows back out through the right depressurizing process. The summation of the work added to the system plus the heat added minus the heat removed yields the net mechanical work generated by the system.

Processes

The processes are described by:

- Process 0–1, a mass of air is drawn into piston/cylinder arrangement at constant pressure.
- Process 1–2 is an adiabatic (isentropic) compression of the charge as the piston moves from bottom dead centre (*BDC*) to top dead centre (*TDC*).
- Process 2–3 is a constant-volume heat transfer to the working gas from an external source while the piston is at top dead centre. This process is intended to represent the ignition of the fuel-air mixture and the subsequent rapid burning.
- Process 3–4 is an adiabatic (isentropic) expansion (power stroke).
- Process 4–1 completes the cycle by a constant-volume process in which heat is rejected from the air while the piston is at bottom dead centre.
- Process 1–0, the mass of air is released to the atmosphere in a constant pressure process.

The Otto cycle consists of isentropic compression, heat addition at constant volume, isentropic expansion, and rejection of heat at constant volume. In the case of a four-stroke Otto cycle, technically there are two additional processes: one for the exhaust of waste heat and combustion products at constant pressure (isobaric), and one for the intake of cool oxygen-rich air also at constant pressure; however, these are often omitted in a simplified analysis. Even though those two processes are critical to the functioning of a real engine, wherein the details of heat transfer and combustion chemistry are relevant, for the simplified analysis of the thermodynamic cycle, it is more convenient to assume that all of the waste-heat is removed during a single volume change.

Processes

The system is defined to be the mass of air that is drawn from the atmosphere into the cylinder, compressed by the piston, heated by the spark ignition of the added fuel, allowed to expand as it pushes on the piston, and finally exhausted back into the atmosphere. The mass of air is followed as its volume, pressure and temperature change during the various thermodynamic steps. As the piston is capable of moving along the cylinder, the volume of the air changes with its position in the cylinder. The compression and expansion processes induced on the gas by the movement of the piston are idealised as reversible, i.e., no useful work is lost through turbulence or friction and no heat is transferred to or from the gas during those two processes. Energy is added to the air by the combustion of fuel. Useful work is extracted by the expansion of the gas in the cylinder. After the expansion is completed in the cylinder, the remaining heat is extracted and finally the gas is exhausted to the environment. Useful mechanical work is produced during the expansion process and some of that used to compress the air mass of the next cycle. The useful mechanical work produced minus that used for the compression process is the net work gained and that can be used for propulsion or for driving other machines. Alternatively the useful work gained is the difference between the heat added and the heat removed.

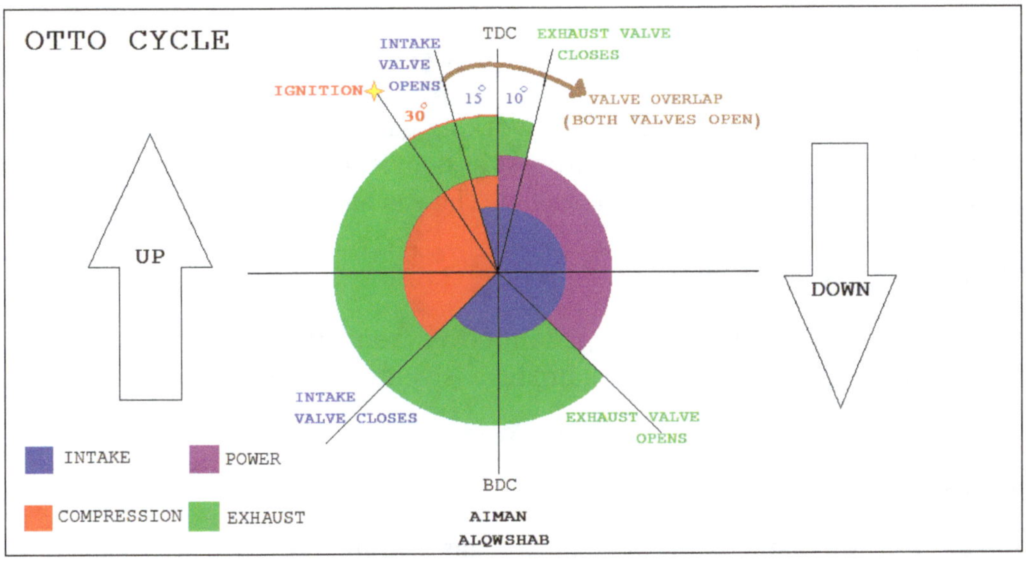

Process 0–1 Intake Stroke (Green Arrow)

A mass of air (working fluid) is drawn into the cylinder, from 0 to 1, at atmospheric pressure (constant pressure) through the open intake valve, while the exhaust valve is closed during this process. The intake valve closes at point 1.

Process 1–2 Compression Stroke (*B* on Diagram)

Piston moves from crank end (BDC, bottom dead centre and maximum volume) to cylinder head end (*TDC*, top dead centre and minimum volume) as the working gas with initial state 1 is compressed isentropically to state point 2, through compression ratio (V_1/V_2). Mechanically this is the isentropic compression of the air/fuel mixture in the cylinder, also known as the compression stroke. This isentropic process assumes that no mechanical energy is lost due to friction and no heat is transferred to or from the gas, hence the process is reversible. The compression process requires that mechanical work be added to the working gas. Generally the compression ratio is around 9–10:1 (V_1:V_2) for a typical engine.

Process 2–3 Ignition Phase (*C* on Diagram)

The piston is momentarily at rest at *TDC*. During this instant, which is known as the ignition phase, the air/fuel mixture remains in a small volume at the top of the compression stroke. Heat is added to the working fluid by the combustion of the injected fuel, with the volume essentially being held constant. The pressure rises and the ratio (P_3 / P_2) is called the "explosion ratio".

Process 3–4 Expansion Stroke (*D* on Diagram)

The increased high pressure exerts a force on the piston and pushes it towards the *BDC*. Expansion of working fluid takes place isentropically and work is done by the system on the piston. The volume ratio V_4 / V_3 is called the "isentropic expansion ratio". (For the Otto cycle is the same as the compression ratio V_1 / V_2). Mechanically this is the expansion of the hot gaseous mixture in the cylinder known as expansion (power) stroke.

Process 4–1 Idealized Heat Rejection (*A* on Diagram)

The piston is momentarily at rest at *BDC*. The working gas pressure drops instantaneously from point 4 to point 1 during a constant volume process as heat is removed to an idealized external sink that is brought into contact with the cylinder head. The gas has returned to state 1.

Process 1–0 Exhaust Stroke

The exhaust valve opens at point 1. As the piston moves from "BDC" (point 1) to "TDC" (point 0) with the exhaust valve opened, the gaseous mixture is vented to the atmosphere and the process starts anew.

Cycle Analysis

In the processes 1–2 the piston does work on the gas and in process 3–4 the gas does work on the piston during those isentropic compression and expansion processes, respectively. Processes 2–3 and 4–1 are isochoric processes; heat is transferred into the system from 2—3 and out of the system from 4—1 but no work is done on the system or extracted from the system during those processes. No work is done during an isochoric (constant volume) process because addition or removal of work from a system requires the movement of the boundaries of the system; hence, as the cylinder volume does not change, no shaft work is added to or removed from the system.

Four different equations are used to describe those four processes. A simplification is made by assuming changes of the kinetic and potential energy that take place in the system (mass of gas) can be neglected and then applying the first law of thermodynamics (energy conservation) to the mass of gas as it changes state as characterized by the gas's temperature, pressure, and volume.

During a complete cycle, the gas returns to its original state of temperature, pressure and volume, hence the net internal energy change of the system (gas) is zero. As a result, the energy (heat or work) added to the system must be offset by energy (heat or work) that leaves the system. In the analysis of thermodynamic systems, the convention is to account energy that enters the system as positive and energy that leaves the system is accounted as negative.

a. During a complete cycle, the net change of energy of the system is zero:

$$\Delta E = E_{in} - E_{out} = 0$$

The above states that the system (the mass of gas) returns to the original thermodynamic state it was in at the start of the cycle.

Where E_{in} is energy added to the system from 1–2–3 and E_{out} is energy is removed from 3–4–1. In terms of work and heat added to the system.

b. $W_{1-2} + Q_{2-3} + W_{3-4} + Q_{4-1} = 0$

Each term of the equation can be expressed in terms of the internal energy of the gas at each point in the process:

$$W_{1-2} = U_2 - U_1$$

$$Q_{2-3} = U_3 - U_2$$

$$W_{3-4} = U_4 - U_3$$

$$Q_{4-1} = U_1 - U_4$$

The energy balance equation becomes:

$$W_{1-2} + Q_{2-3} + W_{3-4} + Q_{4-1} = (U_2 - U_1) + (U_3 - U_2) + (U_4 - U_3) + (U_1 - U_4) = 0$$

To illustrate the example we choose some values to the points in the illustration:

$$U_1 = 1$$

$$U_2 = 5$$

$$U_3 = 9$$

$$U_4 = 4$$

These values are arbitrarily but rationally selected. The work and heat terms can then be calculated.

The energy added to the system as work during the compression from 1 to 2 is:

$$(U_2 - U_1) = (5 - 1) = 4$$

The energy added to the system as heat from point 2 to 3 is:

$$(U_3 - U_2) = (9 - 5) = 4$$

The energy removed from the system as work during the expansion from 3 to 4 is:

$$(U_4 - U_3) = (4 - 9) = -5$$

The energy removed from the system as heat from point 4 to 1 is:

$$(U_1 - U_4) = (1 - 4) = -3$$

The energy balance is:

$$\Delta E = +4 + 4 - 5 - 3 = 0$$

Note that energy added to the system is counted as positive and energy leaving the system is counted as negative and the summation is zero as expected for a complete cycle that returns the system to its original state.

From the energy balance the work out of the system is:

$$\sum \text{Work} = W_{1-2} + W_{3-4} = (U_2 - U_1) + (U_4 - U_3) = 4 - 5 = -1$$

The net energy out of the system as work is -1, meaning the system has produced one net unit of energy that leaves the system in the form of work.

The net heat out of the system is:

$$\sum \text{Heat} = Q_{2-3} + Q_{4-1} = (U_3 - U_2) + (U_1 - U_4) = 4 - 3 = 1$$

As energy added to the system as heat is positive. From the above it appears as if the system gained one unit of heat. This matches the energy produced by the system as work out of the system.

Thermal efficiency is the quotient of the net work from the system, to the heat added to system.

$$\eta = \frac{W_{1-2} + W_{3-4}}{Q_{2-3}} = \frac{(U_2 - U_1) + (U_4 - U_3)}{(U_3 - U_2)}$$

$$\eta = 1 + \frac{U_1 - U_4}{(U_3 - U_2)} = 1 + \frac{(1-4)}{(9-5)} = 0.25$$

Alternatively, thermal efficiency can be derived by strictly heat added and heat rejected.

$$\eta = \frac{Q_{2-3} + Q_{4-1}}{Q_{2-3}} = 1 + \frac{(U_1 - U_4)}{(U_3 - U_2)}$$

Supplying the fictitious values:

$$\eta = 1 + \frac{1-4}{9-5} = 1 + \frac{-3}{4} = 0.25$$

In the Otto cycle, there is no heat transfer during the process 1–2 and 3–4 as they are isentropic processes. Heat is supplied only during the constant volume processes 2–3 and heat is rejected only during the constant volume processes 4–1.

The above values are absolute values that might, for instance, have units of joules (assuming the MKS system of units are to be used) and would be of use for a particular engine with particular dimensions. In the study of thermodynamic systems the extensive quantities such as energy, volume, or entropy (versus intensive quantities of temperature and pressure) are placed on a unit mass basis, and so too are the calculations, making those more general and therefore of more general use. Hence, each term involving an extensive quantity could be divided by the mass, giving the terms units of joules/kg (specific energy), meters^3/kg (specific volume), or joules/(kelvin·kg) (specific entropy, heat capacity) etc. and would be represented using lower case letters, u, v, s, etc.

The specific heats are particularly useful for thermodynamic calculations involving the ideal gas model:

$$C_v = \left(\frac{\delta u}{\delta T}\right)_v$$

Rearranging yields:

$$\delta u = (C_v)(\delta T)$$

Inserting the specific heat equation into the thermal efficiency equation yields:

$$\eta = 1 - \left(\frac{C_v(T_4 - T_1)}{C_v(T_3 - T_2)}\right)$$

Upon rearrangement:

$$\eta = 1 - \left(\frac{T_1}{T_2}\right)\left(\frac{T_4 / T_1 - 1}{T_3 / T_2 - 1}\right)$$

Next, noting from the diagrams $T_4 / T_1 = T_3 / T_2$, thus both of these can be omitted. The equation then reduces to:

$$\eta = 1 - \left(\frac{T_1}{T_2}\right)$$

Since the Otto cycle uses isentropic processes during the compression (process 1 to 2) and expansion (process 3 to 4) the isentropic equations of ideal gases and the constant pressure/volume relations can be used to yield following equations.

$$\left(\frac{T_2}{T_1}\right)=\left(\frac{p_2}{p_1}\right)^{(\gamma-1)/\gamma}$$

$$\left(\frac{T_2}{T_1}\right)=\left(\frac{V_1}{V_2}\right)^{(\gamma-1)}$$

where,

$$\gamma=\left(\frac{C_p}{C_v}\right)$$

γ is the specific heat ratio.

The derivation of the previous equations are found by solving these four equations respectively (where R is the specific gas constant):

$$C_p\ln\left(\frac{V_1}{V_2}\right)-R\ln\left(\frac{p_2}{p_1}\right)=0$$

$$C_v\ln\left(\frac{T_2}{T_1}\right)-R\ln\left(\frac{V_2}{V_1}\right)=0$$

$$C_p=\left(\frac{\gamma R}{\gamma-1}\right)$$

$$C_v=\left(\frac{R}{\gamma-1}\right)$$

Further simplifying above equation, where r is the compression ratio (V_1/V_2):

$$\left(\frac{T_2}{T_1}\right)=\left(\frac{V_1}{V_2}\right)^{(\gamma-1)}=r^{(\gamma-1)}$$

From inverting equation $\left(\frac{T_2}{T_1}\right)=\left(\frac{V_1}{V_2}\right)^{(\gamma-1)}$ and inserting it into equation $\eta=1-\left(\frac{T_1}{T_2}\right)$ the final thermal efficiency can be expressed as: name="Fundamentals" />.

$$\eta=1-\left(\frac{1}{r^{(\gamma-1)}}\right)$$

From analyzing equation above it is evident that the Otto cycle efficiency depends directly upon the compression ratio r. Since the γ for air is 1.4, an increase in η will produce an increase in. However, the γ for combustion products of the fuel/air mixture is often taken at approximately 1.3. The foregoing discussion implies that it is more efficient to have a high compression ratio. The standard ratio is approximately 10:1 for typical automobiles. Usually this does not increase much because of the possibility of autoignition, or “knock”, which places an upper limit on the compression ratio. During the compression process 1–2 the temperature rises, therefore an increase in the compression ratio causes an increase in temperature. Autoignition occurs when the temperature of the fuel/air mixture becomes too high before it is ignited by the flame front. The compression stroke is intended to compress the products before the flame ignites the mixture. If the compression ratio is increased, the mixture may auto-ignite before the compression stroke is complete, leading to “engine knocking”. This can damage engine components and will decrease the brake horsepower of the engine.

Power

The power produced by the Otto cycle is the energy developed per unit of time. The Otto engines are called four-stroke engines. The intake stroke and compression stroke require one rotation of the engine crankshaft. The power stroke and exhaust stroke require another rotation. For two rotations there is one work generating stroke.

From the above cycle analysis the net work produced by the system:

$$\sum \text{Work} = W_{1-2} + W_{3-4} = (U_2 - U_1) + (U_4 - U_3) = +4 - 5 = -1$$

(again, using the sign convention, the minus sign implies energy is leaving the system as work)

If the units used were MKS the cycle would have produced one joule of energy in the form of work. For an engine of a particular displacement, such as one liter, the mass of gas of the system can be calculated assuming the engine is operating at standard temperature (20 °C) and pressure (1 atm). Using the Universal Gas Law the mass of one liter of gas is at room temperature and sea level pressure:

$$M = \frac{PV}{RT}$$

Where, V=0.001 m³, R=0.286 kJ/(kg·K), T=293 K, P=101.3 kN/m².

M=0.00121 kg.

At an engine speed of 3000 RPM there are 1500 work-strokes/minute or 25 work-strokes/second.

$$\sum \text{Work} = 1\text{J}/(\text{kg}\cdot\text{stroke}) \times 0.00121\text{kg} = 0.00121\text{J}/\text{stroke}.$$

Power is 25 times that since there are 25 work-strokes/second:

$$P = 25 \times 0.00121 = 0.0303\text{J}/\text{s or W}$$

If the engine is multi-cylinder, the result would be multiplied by that factor. If each cylinder is of a different liter displacement, the results would also be multiplied by that factor. These results are the product of the values of the internal energy that were assumed for the four states of the system at the end each of the four strokes (two rotations). They were selected only for the sake of illustration, and are obviously of low value. Substitution of actual values from an actual engine would produce results closer to that of the engine. Whose results would be higher than the actual engine as there are many simplifying assumptions made in the analysis that overlook inefficiencies. Such results would overestimate the power output.

Increasing Power and Efficiency

The difference between the exhaust and intake pressures and temperatures means that some increase in efficiency can be gained by use of a turbocharger, removing from the exhaust flow some part of the remaining energy and transferring that to the intake flow to increase the intake pressure. A gas turbine can extract useful work energy from the exhaust stream and that can then be used to pressurize the intake air. The pressure and temperature of the exhausting gases would be reduced as they expand through the gas turbine and that work is then applied to the intake gas stream, increasing its pressure and temperature. The transfer of energy amounts to an efficiency improvement and the resulting power density of the engine is also improved. The intake air is typically cooled so as to reduce its volume as the work produced per stroke is a direct function of the amount of mass taken into the cylinder; denser air will produce more work per cycle. Practically speaking the intake air mass temperature must also be reduced to prevent premature ignition in a petrol fueled engine; hence, an intercooler is used to remove some energy as heat and so reduce the intake temperature. Such a scheme both increases the engine's efficiency and power.

The application of a supercharger driven by the crankshaft does increase the power output (power density) but does not increase efficiency as it uses some of the net work produced by the engine to pressurize the intake air and fails to extract otherwise wasted energy associated with the flow of exhaust at high temperature and a pressure to the ambient.

Diesel Cycle

The Diesel cycle is a combustion process of a reciprocating internal combustion engine. In it, fuel is ignited by heat generated during the compression of air in the combustion chamber, into which fuel is then injected. This is in contrast to igniting the fuel-air mixture with a spark plug as in the Otto cycle (four-stroke/petrol) engine. Diesel engines are used in aircraft, automobiles, power generation, diesel-electric locomotives, and both surface ships and submarines.

The Diesel cycle is assumed to have constant pressure during the initial part of the combustion phase. This is an idealized mathematical model: real physical diesels do have an increase in pressure during this period, but it is less pronounced than in the Otto cycle. In contrast, the idealized Otto cycle of a gasoline engine approximates a constant volume process during that phase.

Idealized Diesel Cycle

The image shows a p-V diagram for the ideal Diesel cycle; where p is pressure and V the volume or v the specific volume if the process is placed on a unit mass basis. The *idealized* Diesel cycle assumes an ideal gas and ignores combustion chemistry, exhaust and recharge procedures and simply follows four distinct processes:

- 1→2: Isentropic compression of the fluid (blue).

- 2→3: Reversible constant pressure heating (red).
- 3→4: Isentropic expansion (yellow).
- 4→1: Reversible constant volume cooling (green).

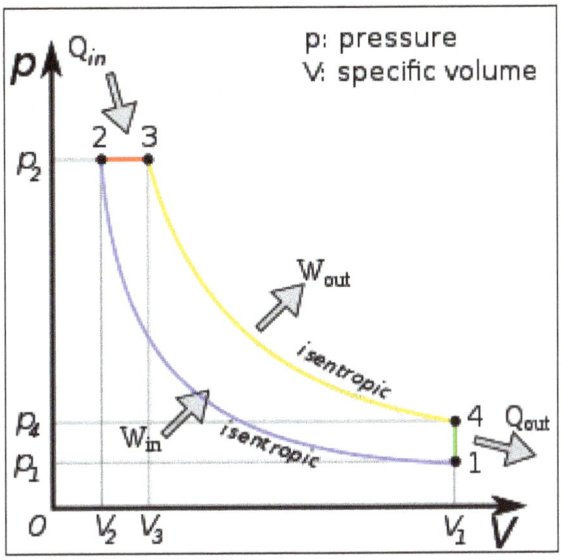

p-V Diagram for the ideal Diesel cycle. The cycle follows the numbers 1-4 in clockwise direction.

The Diesel engine is a heat engine: it converts heat into work. During the bottom isentropic processes (blue), energy is transferred into the system in the form of work W_{in}, but by definition (isentropic) no energy is transferred into or out of the system in the form of heat. During the constant pressure (red, isobaric) process, energy enters the system as heat Q_{in}. During the top isentropic processes (yellow), energy is transferred out of the system in the form of W_{out}, but by definition (isentropic) no energy is transferred into or out of the system in the form of heat. During the constant volume (green, isochoric) process, some of energy flows out of the system as heat through the right depressurizing process Q_{out}. The work that leaves the system is equal to the work that enters the system plus the difference between the heat added to the system and the heat that leaves the system; in other words, net gain of work is equal to the difference between the heat added to the system and the heat that leaves the system.

- Work in (W_{in}) is done by the piston compressing the air (system).
- Heat in (Q_{in}) is done by the combustion of the fuel.
- Work out (W_{out}) is done by the working fluid expanding and pushing a piston (this produces usable work).
- Heat out (Q_{out}) is done by venting the air.
- Net work produced = Q_{in} - Q_{out}.

The net work produced is also represented by the area enclosed by the cycle on the P-V diagram. The net work is produced per cycle and is also called the useful work, as it can be turned to other useful types of energy and propel a vehicle (kinetic energy) or produce electrical energy. The summation of many such cycles per unit of time is called the developed power. The W_{out} is also called the gross work, some of which is used in the next cycle of the engine to compress the next charge of air.

Maximum Thermal Efficiency

The maximum thermal efficiency of a Diesel cycle is dependent on the compression ratio and the cut-off ratio. It has the following formula under cold air standard analysis:

$$\eta_{th} = 1 - \frac{1}{r^{\gamma-1}}\left(\frac{\alpha^{\gamma}-1}{\gamma(\alpha-1)}\right)$$

where,

- η_{th} is thermal efficiency
- α is the cut-off ratio $\frac{V_3}{V_2}$ (ratio between the end and start volume for the combustion phase)
- r is the compression ratio $\frac{V_1}{V_2}$
- γ is ratio of specific heats (C_p/C_v)

The cut-off ratio can be expressed in terms of temperature as shown below:

$$\frac{T_2}{T_1} = \left(\frac{V_1}{V_2}\right)^{\gamma-1} = r^{\gamma-1}$$

$$T_2 = T_1 r^{\gamma-1}$$

$$\frac{V_3}{V_2} = \frac{T_3}{T_2}$$

$$\alpha = \left(\frac{T_3}{T_1}\right)\left(\frac{1}{r^{\gamma-1}}\right)$$

T_3 can be approximated to the flame temperature of the fuel used. The flame temperature can be approximated to the adiabatic flame temperature of the fuel with corresponding air-to-fuel ratio and compression pressure, p_3. T_1 can be approximated to the inlet air temperature.

This formula only gives the ideal thermal efficiency. The actual thermal efficiency will be significantly lower due to heat and friction losses. The formula is more complex than the Otto cycle (petrol/gasoline engine) relation that has the following formula:

$$\eta_{otto,th} = 1 - \frac{1}{r^{\gamma-1}}$$

The additional complexity for the Diesel formula comes around since the heat addition is at constant pressure and the heat rejection is at constant volume. The Otto cycle by comparison has both the heat addition and rejection at constant volume.

Comparing Efficiency to Otto Cycle

Comparing the two formulae it can be seen that for a given compression ratio (r), the *ideal* Otto cycle will be more efficient. However, a *real* diesel engine will be more efficient overall since it will have the ability to operate at higher compression ratios. If a petrol engine were to have the same compression ratio, then knocking (self-ignition) would occur and this would severely reduce the efficiency, whereas in a diesel engine, the self ignition is the desired behavior. Additionally, both of these cycles are only idealizations, and the actual behavior does not divide as clearly or sharply. Furthermore, the ideal Otto cycle formula stated above does not include throttling losses, which do not apply to diesel engines.

Applications

Diesel Engines

Diesel engines have the lowest specific fuel consumption of any large internal combustion engine employing a single cycle, 0.26 lb/hp·h (0.16 kg/kWh) for very large marine engines (combined cycle power plants are more efficient, but employ two engines rather than one). Two-stroke diesels with high pressure forced induction, particularly turbocharging, make up a large percentage of the very largest diesel engines.

In North America, diesel engines are primarily used in large trucks, where the low-stress, high-efficiency cycle leads to much longer engine life and lower operational costs. These advantages also make the diesel engine ideal for use in the heavy-haul railroad and earthmoving environments.

Internal Combustion Engines without Spark Plugs

Many model airplanes use very simple "glow" and "diesel" engines. Glow engines use glow plugs. "Diesel" model airplane engines have variable compression ratios. Both types depend on special fuels.

Some 19th-century or earlier experimental engines used external flames, exposed by valves, for ignition, but this becomes less attractive with increasing compression. (It was the research of Nicolas Léonard Sadi Carnot that established the thermodynamic value of compression.) A historical implication of this is that the diesel engine could have been invented without the aid of electricity.

Atkinson Cycle

Atkinson cycle is an ideal cycle for Otto engine exhausting to a gas turbine. In this cycle the isentropic expansion (3-4) of an Otto cycle (1-2-3-4) is further allowed to proceed to the lowest cycle pressure so as to increase the work output. With this modification the cycle is known as Atkinson cycle. The cycle is shown on p-v and T-s diagrams in figure. Processes involved are:

- Process 1-2: Reversible adiabatic compression (v_1 to v_2).

- Process 2-3: Constant volume heat addition.
- Process 3-4: Reversible adiabatic expansion (v_3 to v_4).
- Process 4-1: Constant pressure heat rejection.

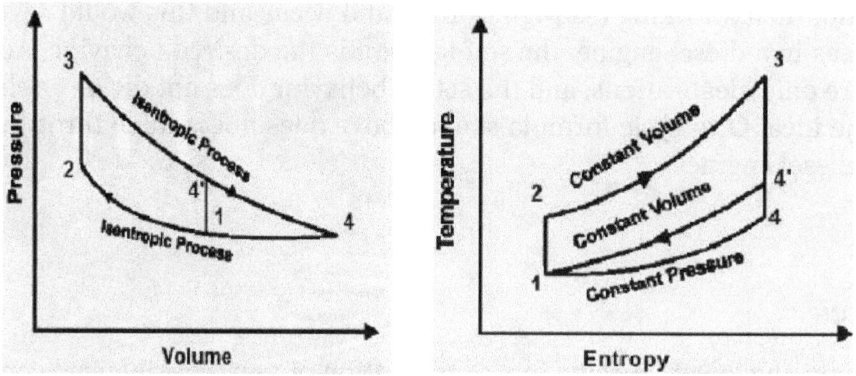

Atkinson cycle on p-v and T-s diagrams.

Thermal Efficiency:

$$\text{Heat supplied} = C_v (T_3 - T_2)$$

$$\text{Heat rejected} = C_P (T_4 - T_1)$$

$$\text{Net workdone} = C_v (T_3 - T_2) - C_P (T_4 - T_1)$$

$$\eta_{th} = \frac{C_v (T_3 - T_2) - C_P (T_4 - T_1)}{C_v (T_3 - T_2)}$$

$$= 1 - \frac{\gamma(T_4 - T_1)}{(T_3 - T_2)}$$

$$\text{Let } r = \frac{v_1}{v_2} = CR$$

$$T_2 = T_1 r^{\gamma - 1}$$

$$\frac{T_3}{T_2} = \frac{P_3}{P_2} = r_P = \text{Pressure ratio}$$

$$T_3 = T_2 r_p = T_1 r^{\gamma - 1} r_p$$

$$\frac{T_3}{T_4} = \left(\frac{P_3}{P_4}\right)^{\frac{\gamma-1}{\gamma}} = \left(\frac{P_3}{P_1}\right)^{\frac{\gamma-1}{\gamma}} = \left(\frac{P_3}{P_2} \cdot \frac{P_2}{P_1}\right)^{\frac{\gamma-1}{\gamma}} = \left(r_p \cdot \frac{P_2}{P_1}\right)^{\frac{\gamma-1}{\gamma}} = r_p^{\frac{\gamma-1}{\gamma}} . r^{\gamma-1}$$

since,

$$\frac{P_2}{P_1} = \left(\frac{v_1}{v_2}\right)^{\gamma} = r^{\gamma}$$

and

$$T_4 = \frac{T_3}{r_p^{\frac{\gamma-1}{\gamma}} r^{\gamma-1}} = \frac{T_1 r_p r^{\gamma-1}}{r_p^{\frac{\gamma-1}{\gamma}} r^{\gamma-1}} = T_1 r_p^{\frac{1}{\gamma}}$$

$$\eta_{th} = 1 - \gamma \left[\frac{r_p^{\frac{1}{\gamma}} - 1}{(r_p - 1) r^{\gamma-1}} \right]$$

Ericsson Cycle

The Ericsson cycle consists of two isothermal and two constant pressure processes. The processes are:

- Process 1-2: Reversible isothermal compression.
- Process 2-3: Constant pressure heat addition.
- Process 3-4: Reversible isothermal expansion.
- Process 4-1: Constant pressure heat rejection.

The heat addition and rejection take place at constant pressure as well as isothermal processes. Since the process 2-3 and 3-4 are parallel to each other on the T-s diagram, the net effect is that the heat need to be added only at constant temperature T_3=T_4 and rejected at the constant temperature T_1=T_2. The cycle is shown on p-v and T-s diagrams in figure. The advantage of the Ericsson cycle over the Carnot and Stirling cycles is its smaller pressure ratio for a given ratio of maximum to minimum specific volume with higher mean effective pressure.

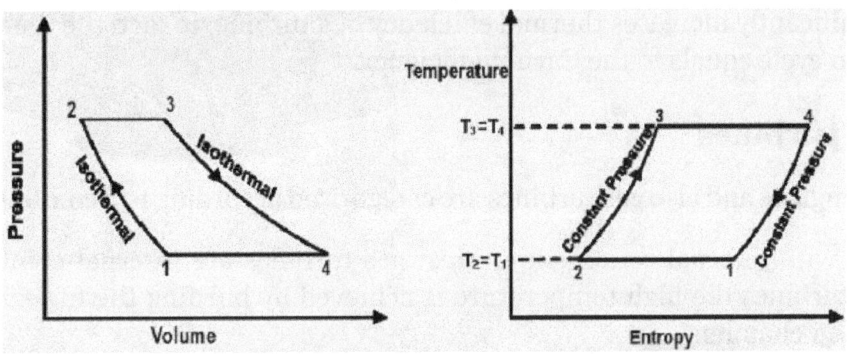

Ericsson cycle on p-v and T-s diagrams.

The thermal efficiency of Ericsson cycle is given by, (derivation is same as that of Stirling cycle),

$$\eta_{th} = \frac{T_H - T_L}{T_H} = \left[1 - \frac{T_L}{T_H} \right]$$

The Ericsson cycle does not find practical application in piston engines but is approached by a gas turbine employing a large number of stages with heat exchangers, insulators and reheaters.

Brayton Cycle

The Brayton cycle describes the workings of a constant-pressure heat engine. It is the one of most common thermodynamic cycles that can be found in gas turbine power plants or in airplanes. In contrast to Carnot cycle, the Brayton cycle does not execute isothermal processes, because these must be performed very slowly. In an ideal Brayton cycle, the system executing the cycle undergoes a series of four processes: two isentropic (reversible adiabatic) processes alternated with two isobaric processes.

Since Carnot's principle states that no engine can be more efficient than a reversible engine (a Carnot heat engine) operating between the same high temperature and low temperature reservoirs, a gas turbine based on the Brayton cycle must have lower efficiency than the Carnot efficiency.

A large single-cycle gas turbine typically produces for example 300 megawatts of electric power and has 35–40% thermal efficiency. Modern Combined Cycle Gas Turbine (CCGT) plants, in which the thermodynamic cycle of consists of two power plant cycles (e.g. the Brayton cycle and the Rankine cycle), can achieve a thermal efficiency of around 55%.

Brayton Cycle vs. Ericsson Cycle

It must be noted the first Ericsson cycle is in fact now called the "Brayton cycle", commonly applied to modern gas turbine engines and airbreathing jet engines. The second Ericsson cycle is similar to the Brayton cycle, but uses external heat and incorporates the use of a regenerator. An ideal Ericsson cycle consists of isothermal compression and expansion processes, combined with isobaric heat regeneration between them. Applying intercooling, heat regeneration and sequential combustion, significantly increases thermal efficiency of a turbine, in fact, the thermal efficiency of the ideal Ericsson cycle equals to the Carnot efficiency.

Types of Gas Turbines

In general, heat engines and also gas turbines are categorized according to a combustion location as:

- Turbines with internal combustion. Most gas turbines are internal combustion engines. In these turbines the high temperature is achieved by burning the fuel-air mixture in the combustion chamber.

- Turbines with external combustion. In these turbines, a heat exchanger is usually used and only clean medium with no combustion products travels through the power turbine. Since the turbine blades are not subjected to combustion products, much lower quality (and therefore cheaper) fuels are able to be used. These turbines have usually lower thermal efficiency than turbines with internal combustion.

Types of Brayton Cycle

Open Brayton Cycle

Since most gas turbines are based on the Brayton cycle with internal combustion (e.g. jet engines), they are based on the open Brayton cycle. In this cycle, air from the ambient atmosphere is compressed to a higher pressure and temperature by the compressor. In the combustion chamber, air is heated further by burning the fuel-air mixture in the air flow. Combustion products and gases expand in the turbine either to near atmospheric pressure (engines producing mechanical energy or electrical energy) or to a pressure required by the jet engines. The open Brayton cycle means that the gases are discharged directly into the atmosphere.

Closed Brayton Cycle

In a closed Brayton cycle working medium (e.g. helium) recirculates in the loop and the gas expelled from the turbine is reintroduced into the compressor. In these turbines, a heat exchanger (external combustion) is usually used and only clean medium with no combustion products travels through the power turbine. The closed Brayton cycle is used, for example, in closed-cycle gas turbine and high-temperature gas cooled reactors.

Reverse Brayton Cycle – Brayton Refrigeration Cycle

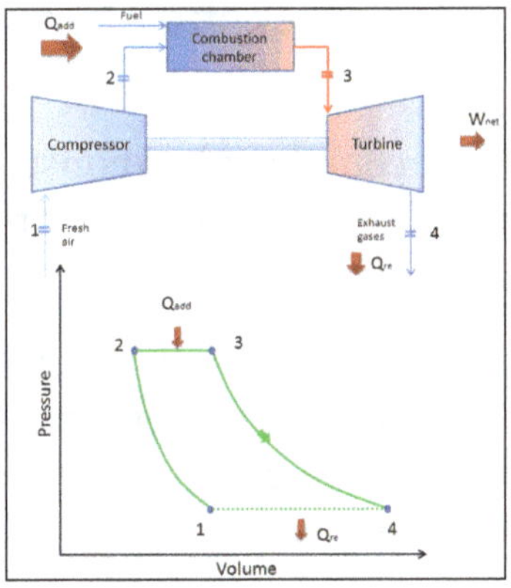

Open Brayton cycle.

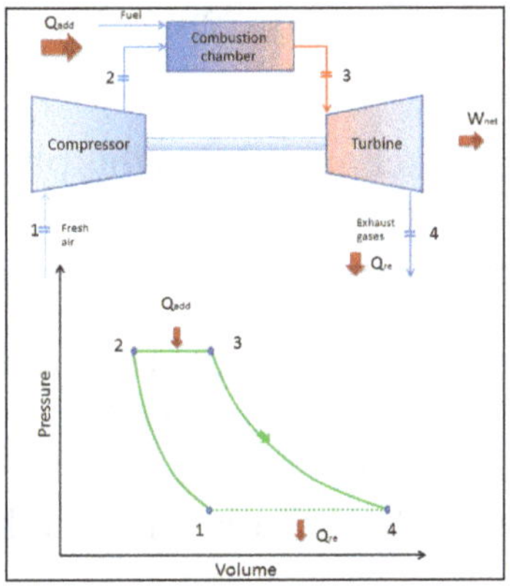

Closed Brayton cycle.

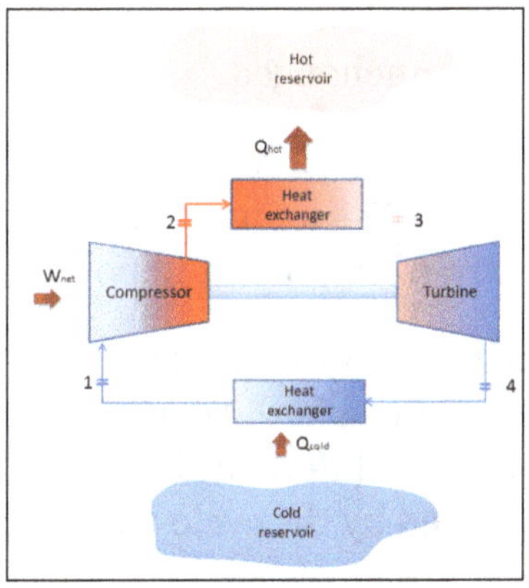

Reverse Brayton cycle.

A Brayton cycle that is driven in reverse direction is known as the reverse Brayton cycle. Its purpose is to move heat from colder to hotter body, rather than produce work. In compliance with the second law of thermodynamics, Heat cannot spontaneously flowfrom cold system to hot system without external work being performed on the system. Heat can flow from colder to hotter body, but only when forced by an external work. This is exactly what refrigerators and heat pumps accomplish. These are driven by electric motors requiring work from their surroundings to operate. One of possible cycles is a reverse Brayton cycle, which is similar to the ordinary Brayton cycle but it is driven in reverse, via net work input. This cycle is also known as the gas refrigeration cycle or Bell Coleman cycle. This type of cycle is widely used in jet aircrafts for air conditioning systems using air from the engine compressors. It is also widely used in the LNG industry where the largest reverse Brayton cycle is for subcooling LNG using 86 MW of power from a gas turbine-driven compressor and nitrogen refrigerant.

Brayton Cycle – Processes

In a closed ideal Brayton cycle, the system executing the cycle undergoes a series of four processes: Two isentropic (reversible adiabatic) processes alternated with two isobaric processes.

- Isentropic compression (compression in a compressor) – The working gas (e.g. helium) is compressed adiabatically from state 1 to state 2 by the compressor (usually an axial-flow compressor). The surroundings do work on the gas, increasing its internal energy (temperature) and compressing it (increasing its pressure). On the other hand the entropy remains unchanged. The work required for the compressor is given by $W_C = H_2 - H_1$.
- Isobaric heat addition (in a heat exchanger) – In this phase (between state 2 and state 3) there is a constant-pressure heat transfer to the gas from an external source, since the chamber is open to flow in and out. In an open ideal Brayton cycle, the compressed air then runs through a combustion chamber, where fuel is burned and air or another medium is heated (2 → 3). It is a constant-pressure process, since the chamber is open to flow in and out. The net heat added is given by $Q_{add} = H_3 - H_2$.
- Isentropic expansion (expansion in a turbine) – The compressed and heated gas expands adiabatically from state 3 to state 4 in a turbine. The gas does work on the surroundings (blades of the turbine) and loses an amount of internal energy equal to the work that leaves the system. The work done by turbine is given by $W_T = H_4 - H_3$. Again the entropy remains unchanged.
- Isobaric heat rejection (in a heat exchanger) – In this phase the cycle completes by a constant-pressure process in which heat is rejected from the gas. The working gas temperature drops from point 4 to point 1. The net heat rejected is given by $Q_{re} = H_4 - H_1$.

During a Brayton cycle, work is done on the gas by the compressor between states 1 and 2 (isentropic compression). Work is done by the gas in the turbine between stages 3 and 4 (isentropic expansion). The difference between the work done by the gas and the work done on the gas is the net work produced by the cycle and it corresponds to the area enclosed by the cycle curve.

As can be seen, it is convenient to use enthalpy or specific enthalpy and to express the first law in terms of enthalpy in analysis of this thermodynamic cycle. This form of the law simplifies the description of energy transfer. At constant pressure, the enthalpy change equals the energy transferred from the environment through heating:

Isobaric process ($Vdp = 0$):

$$dH = dQ \rightarrow Q = H_2 - H_1$$

At constant entropy, i.e. in isentropic process, the enthalpy change equals the flow process work done on or by the system:

Isentropic process ($dQ = 0$):

$$dH = Vdp \rightarrow W = H_2 - H_1$$

Isentropic Process

An isentropic process is a thermodynamic process, in which the entropy of the fluid or gas remains constant. It means the isentropic process is a special case of an adiabatic process in which there is no transfer of heat or matter. It is a reversible adiabatic process. The assumption of no heat transfer is very important, since we can use the adiabatic approximation only in very rapid processes.

Isentropic Process and the First Law

For a closed system, we can write the first law of thermodynamics in terms of enthalpy:

$$dH = dQ + Vdp$$

Or

$$dH = TdS + Vdp$$

Isentropic process ($dQ = 0$):

$$dH = Vdp \rightarrow W = H_2 - H_1 \rightarrow H_2 - H_1 = C_p(T_2 - T_1) \text{ (for ideal gas)}$$

Isentropic Process of the Ideal Gas

The isentropic process (a special case of adiabatic process) can be expressed with the ideal gas law as:

$$pV^k = \text{constant}$$

Or

$$p_1V_1^k = p_2V_2^k$$

in which $k = c_p / c_v$ is the ratio of the specific heats (or heat capacities) for the gas. One for constant pressure (c_p) and one for constant volume (c_v). Note that, this ratio $k = c_p / c_v$ is a factor in determining the speed of sound in a gas and other adiabatic processes.

Isobaric Process

An isobaric process is a thermodynamic process, in which the pressure of the system remains constant (p = const). The heat transfer into or out of the system does work, but also changes the internal energy of the system.

Since there are changes in internal energy (dU) and changes in system volume (ΔV), engineers often use the enthalpy of the system, which is defined as:

$$H = U + pV$$

Isobaric Process and the First Law

The classical form of the first law of thermodynamics is the following equation:

$$dU = dQ - dW$$

In this equation dW is equal to $dW = pdV$ and is known as the boundary work. In an isobaric process and the ideal gas, part of heat added to the system will be used to do work and part of heat added will increase the internal energy (increase the temperature). Therefore it is convenient to use the enthalpy instead of the internal energy.

Isobaric process ($Vdp = 0$):

$$dH = dQ \quad \rightarrow Q = H_2 - H_1$$

At constant entropy, i.e. in isentropic process, the enthalpy change equals the flow process work done on or by the system.

Isobaric Process of the Ideal Gas

The isobaric process can be expressed with the ideal gas law as:

$$\frac{V}{T} = \text{constant}$$

or

$$\frac{V_1}{T_1} = \frac{V_2}{T_2}$$

On a p-V diagram, the process occurs along a horizontal line (called an isobar) that has the equation p = constant.

First Law	$dQ = TdS = o$ $dU = pdV$ $dH = Vdp$
Ideal gas relation	$pV^x = \text{constant}$
p,V,T relations	$\frac{p_1}{p_2} = \left[\frac{V_2}{V_1}\right]^k = \left[\frac{T_1}{T_2}\right]^{\frac{k}{k-1}}$
Change in Internal energy	$dU = mc_V (T_2 - T_1)$
Change in Enthalpy	$dH = mc_p (T_2 - T_1)$
Heat transfer	O
Pdv work	$W_{dV} = pdV = dU = mc_V (T_2 - T_1)$
Vdp work	$W_{dp} = Vdp = dH = mc_p (T_2 - T_1)$

Isentropic Process – Main Characteristics

First Law	$dU = dQ - pdV$ $dH = dQ - Vdp = dQ$
Ideal gas relation	$\frac{V}{T} = \text{constant}$
p,V,T relations	$\frac{V_1}{T_1} = \frac{V_2}{T_2}$
Change in Internal energy	$dU = mc_V (T_2 - T_1)$
Change in Enthalpy	$dH = mc_p (T_2 - T_1)$
Heat transfer	$dH = mc_p (T_2 - T_1)$
Pdv work	$W_{i \to f} = p(V_f - V_i)$ $W_{i \to f} = nR(T_f - T_i)$
Vdp work	$W_{i \to f} = 0$

Isobaric Process – Main Characteristics

The Brayton cycle is often plotted on a pressure volume diagram (pV diagram) and on a temperature-entropy diagram (Ts diagram).

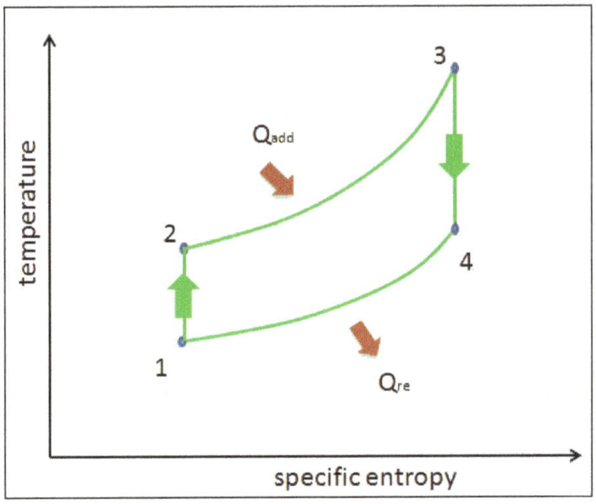

Brayton Cycle – Ts diagram.

When plotted on a pressure volume diagram, the isobaric processes follow the isobaric lines for the gas (the horizontal lines), adiabatic processes move between these horizontal lines and the area bounded by the complete cycle path represents the total work that can be done during one cycle.

The temperature-entropy diagram (Ts diagram) in which the thermodynamic state is specified by a point on a graph with specific entropy (s) as the horizontal axis and absolute temperature (T) as the vertical axis. Ts diagrams are a useful and common tool, particularly because it helps to visualize the heat transfer during a process. For reversible (ideal) processes, the area under the T-s curve of a process is the heat transferred to the system during that process.

Thermal Efficiency of Brayton Cycle

In general the thermal efficiency, η_{th}, of any heat engine is defined as the ratio of the work it does, W, to the heat input at the high temperature, Q_H.

$$\eta_{th} = \frac{W}{Q_H}$$

The thermal efficiency, η_{th}, represents the fraction of heat, Q_H, that is converted to work. Since energy is conserved according to the first law of thermodynamics and energy cannot be be converted to work completely, the heat input, Q_H, must equal the work done, W, plus the heat that must be dissipated as waste heat Q_C into the environment. Therefore we can rewrite the formula for thermal efficiency as:

$$\eta_{th} = \frac{W}{Q_H} = \frac{Q_H - Q_C}{Q_H} = 1 - \frac{Q_C}{Q_H}$$

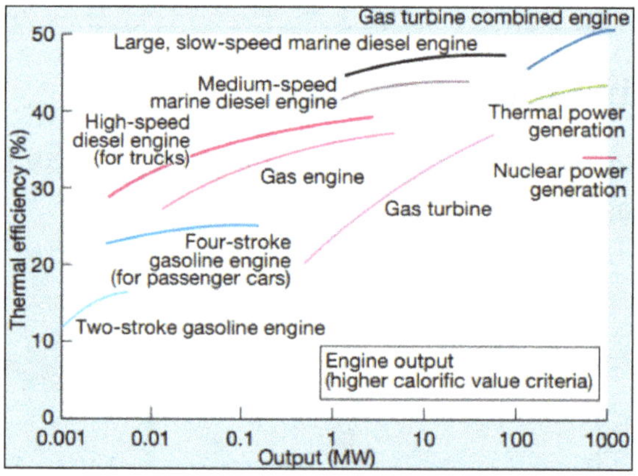

This is very useful formula, but here we express the thermal efficiency using the first law in terms of enthalpy.

To calculate the thermal efficiency of the Brayton cycle (single compressor and single turbine) engineers use the first law of thermodynamics in terms of enthalpy rather than in terms of internal energy.

The first law in terms of enthalpy is:

$$dH = dQ + Vdp$$

In this equation the term *Vdp* is a flow process work. This work, *Vdp*, is used for open flow systems like a turbine or a pump in which there is a "dp", i.e. change in pressure. There are no changes in control volume. As can be seen, this form of the law simplifies the description of energy transfer.

There are expressions in terms of more familiar variables such as temperature and pressure:

$$dH = C_p dT + V(1\text{-}\alpha T)dp$$

Where C_p is the heat capacity at constant pressure and α is the coefficient of (cubic) thermal expansion. For ideal gas $\alpha T = 1$ and therefore:

$$dH = C_p dT$$

At constant pressure, the enthalpy change equals the energy transferred from the environment through heating:

Isobaric process (Vdp = 0):

$$dH = dQ \rightarrow Q = H_3 - H_2 \rightarrow H_3 - H_2 = C_p (T_3 - T_2)$$

At constant entropy, i.e. in isentropic process, the enthalpy change equals the flow process work-done on or by the system:

Isentropic process (dQ = 0):

$$dH = Vdp \rightarrow W = H_4 - H_3 \rightarrow H_4 - H_3 = C_p (T_4 - T_3)$$

The enthalpy can be made into an intensive, or specific, variable by dividing by the mass. Engineers use the specific enthalpy in thermodynamic analysis more than the enthalpy itself.

Now, lets assume the ideal Brayton cycle that describes the workings of a constant pressure heat engine. Modern gas turbine engines and airbreathing jet engines also follow the Brayton cycle. This cycle consist of four thermodynamic processes:

- Brayton Cycle – Ts diagram isentropic compression – ambient air is drawn into the compressor, where it is pressurized (1 → 2). The work required for the compressor is given by $W_C = H_2 - H_1$.
- Isobaric heat addition – The compressed air then runs through a combustion chamber, where fuel is burned and air or another medium is heated (2 → 3). It is a constant-pressure process, since the chamber is open to flow in and out. The net heat added is given by $Q_{add} = H_3 - H_2$.
- Isentropic expansion – The heated, pressurized air then expands on turbine, gives up its energy. The work done by turbine is given by $W_T = H_4 - H_3$.
- Isobaric heat rejection – The residual heat must be rejected in order to close the cycle. The net heat rejected is given by $Q_{re} = H_4 - H_1$.

As can be seen, we can fully describe and calculate such cycles (similarly for Rankine cycle) using enthalpies.

The thermal efficiency of such simple Brayton cycle, for ideal gas and in terms of specific enthalpies can now be expressed in terms of the temperatures:

$$\eta_{th} = \frac{Net\,work}{Heat\,input} = \frac{W_T - W_C}{Q_{in}} = \frac{c_p\left[(T_3 - T_4) - (T_2 - T_1)\right]}{c_p (T_3 - T_2)} = 1 - \frac{(T_4 - T_1)}{(T_3 - T_2)}$$

where,

- W_T the work done by the gas in the turbine.
- W_C the work done on the gas in the compressor.
- c_p is the heat capacity ratio.

Pressure Ratio – Brayton Cycle – Gas Turbine

The thermal efficiency in terms of the compressor pressure ratio ($PR = p_2/p_1$), which is the parameter commonly used:

$$\eta_{th} = 1 - \frac{1}{PR^{k-1/k}}$$

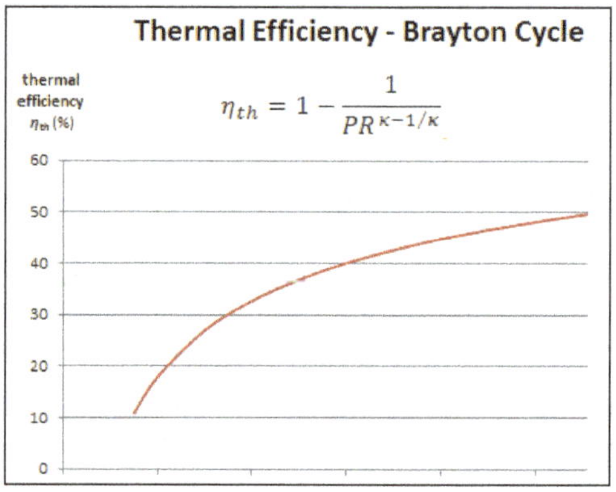

In general, increasing the pressure ratio is the most direct way to increase the overall thermal efficiency of a Brayton cycle, because the cycle approaches the Carnot cycle.

According to Carnot's principle higher efficiencies can be attained by increasing the temperature of the gas.

But there are also limits on the pressure ratios that can be used in the cycle. The highest temperature in the cycle occurs at the end of the combustion process, and it is limited by the maximum temperature that the turbine blades can withstand. As usual, metallurgical considerations (about 1700 K) place an upper limits on thermal efficiency.

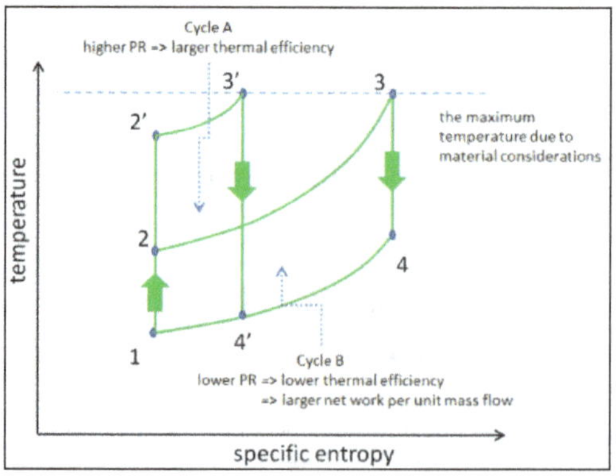

Ideal Brayton cycles with different pressure ratios and the same turbine inlet temperature.

Consider the effect of compressor pressure ratio on thermal efficiency when the turbine inlet temperature is restricted to the maximum allowable temperature. There are two Ts diagrams of Brayton cycles having the same turbine inlet temperature but different compressor pressure ratios in the picture. As can be seen for a fixed-turbine inlet temperature, the net work output per cycle ($W_{net} = W_T - W_C$) decreases with the pressure ratio (Cycle A). But the Cycle A has the greater efficiency.

On the other hand, the Cycle B has a larger net work output per cycle (enclosed area in the diagram) and thus the greater net work developed per unit of mass flow. The work produced by the cycle times a mass flow rate through the cycle is equal to the power output produced by the gas turbine.

Therefore with less work output per cycle (Cycle A), a larger mass flow rate (thus a larger system) is needed to maintain the same power output, which may not be economical. This is the key consideration in design of gas turbine, since here engineers must balance the thermal efficiency and the compactness. In most common designs, the pressure ratio of a gas turbine ranges from about 11 to 16.

Efficiency of Engines in Power Engineering:

- Ocean thermal energy conversion (OTEC). OTEC is very sophisticated heat engine that uses the temperature difference between cooler deep and warmer surface seawaters to run a low-pressure turbine. Since the temperature difference is low, about 20°C, its thermal efficiency is also very low, about 3%.
- In modern nuclear power plants the overall thermal efficiency is about one-third (33%), so 3000 MWth of thermal power from the fission reaction is needed to generate 1000 MWe of electrical power. Higher efficiencies can be attained by increasing the temperature of the steam. But this requires an increase in pressures inside boilers or steam generators. However, metallurgical considerations place an upper limits on such pressures. In comparison to other energy sources the thermal efficiency of 33% is not much. But it must be noted that nuclear power plants are much more complex than fossil fuel power plants and it is much easier to burn fossil fuel, than to generate energy from nuclear fuel.
- Sub-critical fossil fuel power plants, that are operated under critical pressure (i.e. lower than 22.1 MPa), can achieve 36–40% efficiency.
- The supercritical water reactors are considered a promising advancement for nuclear power plants because of its high thermal efficiency (~45 % vs. ~33 % for current LWRs).
- Supercritical fossil fuel power plants, that are operated at supercritical pressure (i.e. greater than 22.1 MPa), have efficiencies around 43%. Most efficient and also very complex coal-fired power plants that are operated at "ultra critical" pressures (i.e. around 30 MPa) and use multiple stage reheat reach about 48% efficiency.
- Modern Combined Cycle Gas Turbine (CCGT) plants, in which the thermodynamic cycle of consists of two power plant cycles (e.g. the Brayton cycle and the Rankine cycle), can achieve a thermal efficiency of around 55%, in contrast to a single cycle steam power plant which is limited to efficiencies of around 35-45%.

Thermal Efficiency Improvement – Brayton Cycle

There are several methods, how can be the thermal efficiency of the Brayton cycle improved. Assuming that the maximum temperature is limited by metallurgical consideration.

Increasing Pressure Ratio

In general, increasing the pressure ratio is the most direct way to increase the overall thermal efficiency of a Brayton cycle, since the thermodynamic efficiency is primarily dependent on the pressure ratio, PR.

$$\eta_{th} = 1 - \frac{1}{PR^{k-1/k}}$$

As was discussed, increasing the pressure ratio increases the compressor discharge temperature. Since the turbine inlet temperature is limited by the maximum temperature that the turbine blades can withstand, the pressure ration influences the heat amount that can be added to the flow. Moreover with an increase of the pressure ratio, the diameter of the compressor blades becomes progressively smaller in higher pressure stages of the compressor. Because the gap between the blades and the engine casing increases in size as a percentage of the compressor blade height as the blades get smaller in diameter, a greater percentage of the compressed air can leak back past the blades in higher pressure stages. This causes a leak back and in result it decreases the isentropic compressor efficiency. Finally, from the formula for the thermal efficiency in term of pressure ratio can be seen, there is smaller gain as the pressure ratio increases (due to the exponent).

Heat Regeneration

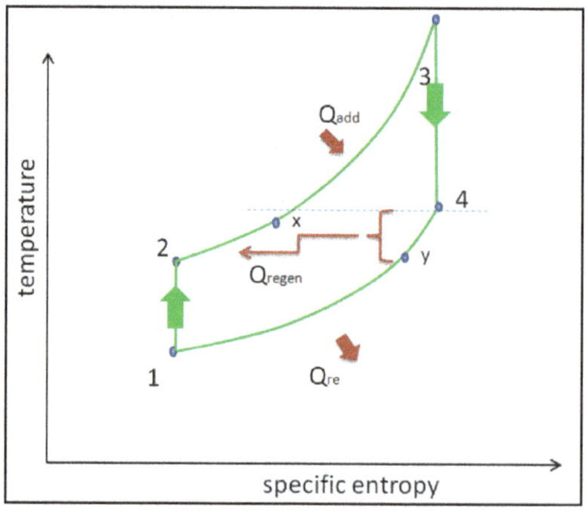

Ts diagram of the Brayton cycle with heat regeneration.

Significant increases in the thermal efficiency of gas turbine power plants can be achieved through reducing the amount of fuel that must be burned in the combustion chamber. This can be done by transferring heat from the turbine exhaust gas, which is normally well above the ambient temperature, to the compressor discharge air flow known as heat regeneration. Especially at a low or moderate pressure ratio, there is a high temperature increase in the combustion chamber and the turbine exhaust gas might still contain significant amount of heat at higher temperature than the compressor outlet gas (after the last compression stage but before the combustor). For this purpose a heat exchanger called a regenerator is used. Sometimes engineers use the term economiser that are heat exchangers intended to reduce energy consumption, especially in case of preheating of a fluid.

This heat regenerator allows the air exiting the compressor to be preheated before it enters the combustion chamber, thereby reducing the amount of fuel that must be burned in the combustor. This form of heat recycling is only possible if the gas turbine is run with low pressure ratio.

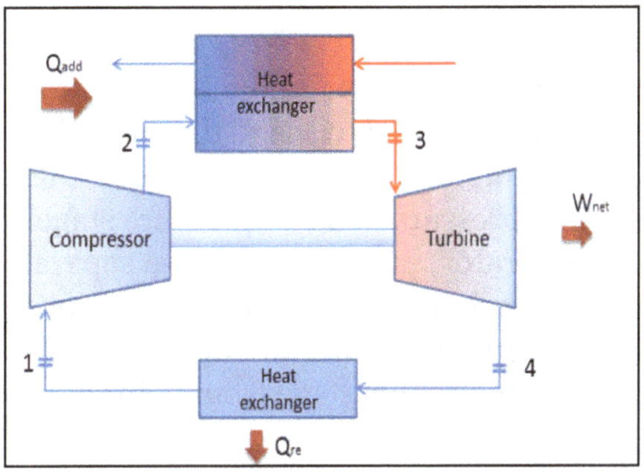

Open Brayton cycle with heat regeneration.

As was stated the temperature difference between turbine outlet and compressor outlet is crucial and determines the amount of heat that can be recovered. In case of negative difference (i.e. $T_2 > T_4$), the heat regeneration is not possible. There are two main ways, how to change this difference:

- To increase the turbine outlet temperature (T_4) through reheat of the flow during expansion phase (i.e. use of a multistage turbine with a reheat combustor or with a reheater).
- To decrease the compressor outlet temperature (T_2) through intercooling of the flow during compression phase (i.e. use of a multistage compressor with an intercooler).

Therefore reheat and intercooling are complementary with heat regeneration. By itself, they would not necessarily increase the thermal efficiency, however, when intercooling or reheat is used in conjunction with heat regeneration, a significant increase in thermal efficiency can be achieved.

It must be noted, transferring heat from the turbine outlet to the compressor inlet would reduce efficiency, as hotter inlet air means more volume, thus more work for the compressor. Engineers must also take into consideration pressure losses generated by the heat exchanger that slightly reduce the power of the gas turbine.

Regeneration vs. Recuperation of Heat

In general, the heat exchangers used in regeneration may be classified as either regeneratorsor recuperators.

- Regenerator is a type of heat exchanger where heat from the hot fluid is intermittently stored in a thermal storage medium before it is transferred to the cold fluid. It has a single flow path in which the hot and cold fluids alternately pass through.
- Recuperator is a type of heat exchanger has separate flow paths for each fluid along their

own passages and heat is transferred through the separating walls. Recuperators (e.g. economisers) are often used in power engineering, to increase the overall efficiency of thermodynamic cycles. For example, in a gas turbine engine. The recuperator transfers some of the waste heat in the exhaust to the compressed air, thus preheating it before entering the combustion chamber. Many recuperators are designed as counterflow heat exchangers.

Reheat - Reheaters

As was discussed, the maximum temperature is limited by metallurgical consideration, but in order to deliver more of the heat at a temperature close to the peak of the cycle the gas can be reheated in a reheater. This involves splitting the turbine, i.e. use of a multistage turbine with a reheat combustor or with a reheater. High pressure and low pressure stages of the turbine may be on the same shaft to drive a common generator, but they will have separate cases. With a reheater, the flow is extracted after a partial expansion (point a), run back through the heat exchanger to heat it back up to the peak temperature (point b), and then passed to the lower pressure stage of the turbine. The expansion then completed in this stage from point b to point 4.

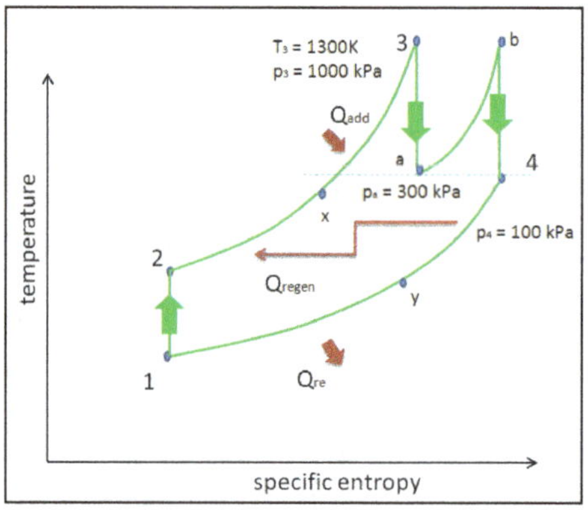

Ts diagram of the Brayton cycle with reheat and regeneration.

With this arrangement the net work per unit of mass flow can be increased. Despite the increase in net work with reheat, the cycle thermal efficiency would not necessarily increase because a greater total heat addition would be required. On the other hand, the temperature at the exit of the turbine (low pressure stage) is higher with reheat than without reheat, so there is the potential for heat regeneration. Therefore reheat and regeneration are complementary, they are used usually together in order to increase the thermal efficiency of gas turbine.

Compression with Intercooling

Significant increases in the thermal efficiency of gas turbine power plants can be achieved also through intercooling. Intercooling can be applied between compressor stages to reduce compression work, W_C, hence increasing overall power of the gas turbine.

For this purpose a heat exchanger known as an intercooler is usually used between stages of a multi-stage compression process. In general, intercoolers are heat exchangers that are used in many applications, including air compressors, air conditioners, refrigerators, and gas turbines. Intercoolers are widely known also in automotive use as a turbocharger or supercharger, but here they increase intake air charge density, hence the power of an engine.

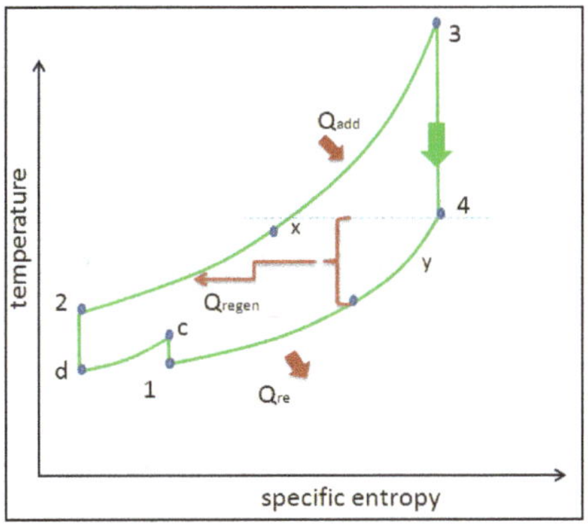

Ts diagram of Brayton cycle with intercooling.

In gas turbine power plant the thermal efficiency is of the highest importance and intercoolingwith heat regeneration are widely used. This involves splitting the compressor, i.e. use of a multistage compressor with an intercooler or intercoolers. High pressure and low pressure stages of the compressor may be on the same shaft even with turbine or a generator, but it is not a rule. With an intercooler, the flow is extracted after a partial compression (point c), run through the heat exchanger (intercooler) to cool it to the ambient temperature (point d), and then passed to the high stage of compressor. The compression is then completed in the second compressor from point d to point 2.

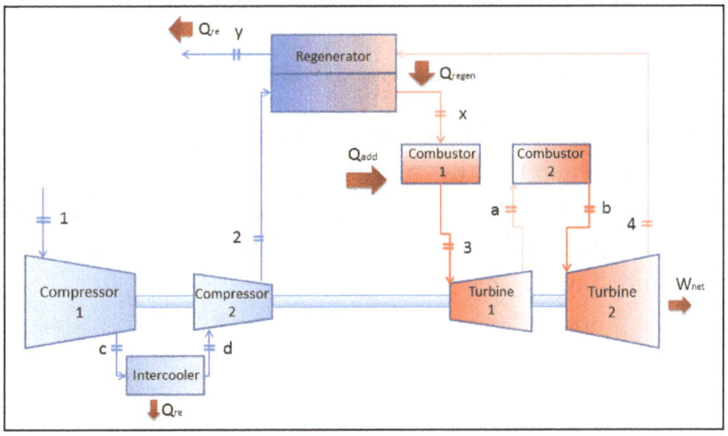

Open Brayton cycle with reheat and intercooling.

With this arrangement the net work per unit of mass flow ($\uparrow W_{net} = W_T - \downarrow W_C$) can be increased by reducing the compression work ($\downarrow W_C$). Despite the increase in net work with intercooling, the cycle thermal efficiency would not necessarily increase because the temperature of the air entering the combustor would be reduced and a greater total heat addition would be required to achieve the desired turbine inlet temperature. On the other hand, the temperature at the exit of the compressor (high pressure stage) is lower with intercooling than without intercooling, so there is the potential for heat regeneration (Q_{regen} increases). Note that, the heat regeneration requires lower compressor outlet temperature than the turbine outlet temperature (simply due to 2nd law) and this temperature difference determines the amount of heat available for heat regeneration.

Therefore reheat and intercooling are complementary with heat regeneration. By itself, they would not necessarily increase the thermal efficiency, however, when intercooling or reheat is used in conjunction with heat regeneration, a significant increase in thermal efficiency can be achieved.

Some large compressors with higher pressure ratio have several stages of compression with inter-cooling between stages. Engineers must also take into consideration pressure lossesgenerated by all heat exchangers that slightly increase compression work. The certain gas turbine design (the number of intercoolers, reheaters and regenerators) is an engineering problem and depends on certain purpose of the gas turbine.

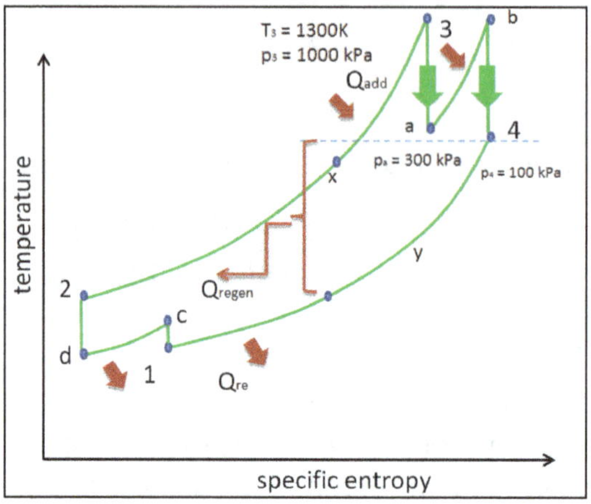

Ts diagram of Brayton cycle with reheat, intercooling and heat regeneration.

Reheat, Intercooling and Regeneration in Brayton Cycle

As was discussed reheat and intercooling are complementary with heat regeneration. By itself, they would not necessarily increase the thermal efficiency, however, when intercooling or reheat is used in conjunction with heat regeneration, a significant increase in thermal efficiency can be achieved and the net work output is also increased. This requires a gas turbine with two stages of compression and two turbine stages.

Ericsson Cycle

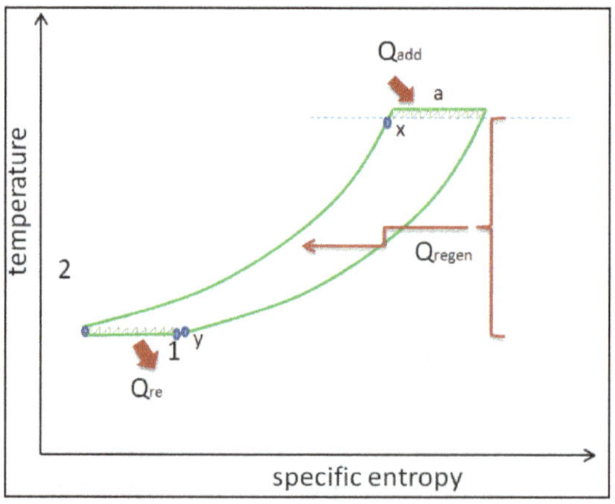

Ericsson cycle – Ts diagram.

The second Ericsson cycle is similar to the Brayton cycle, but uses external heat and incorporates the multiple use of an intercooling and reheat. In fact, it is like a Brayton cycle with an infinite number of reheat and intercooler stages in the cycle. Compared to the Brayton cycle which uses adiabatic compression and expansion, an ideal Ericsson cycle consists of isothermal compression and expansion processes, combined with isobaric heat regeneration between them. Applying intercooling, heat regeneration and sequential combustion significantly increases thermal efficiency of a turbine, in fact, the thermal efficiency of the ideal Ericsson cycle equals to the Carnot efficiency.

Isentropic Efficiency – Turbine, Compressor

Most steady-flow devices (turbines, compressors, nozzles) operate under adiabatic conditions, but they are not truly isentropic but are rather idealized as isentropic for calculation purposes. We define parameters η_T, η_C, η_N, as a ratio of real work done by device to work by device when operated under isentropic conditions (in case of turbine). This ratio is known as the Isentropic Turbine/Compressor/Nozzle Efficiency.

These parameters describe how efficiently a turbine, compressor or nozzle approximates a corresponding isentropic device. This parameter reduces the overall efficiency and work output. For turbines, the value of η_T is typically 0.7 to 0.9 (70–90%).

Isentropic Turbine Efficiency:

$$\eta_T = \frac{\text{Real Turbine Work}}{\text{Isentropic Turbine Work}} = \frac{W_{real}}{W_s} = \frac{h_1 - h_{2r}}{h_1 - h_{2s}}$$

Isentropic Compressor Efficiency:

$$\eta_C = \frac{\text{Isentropic Compressor Work}}{\text{Real Compressor Work}} = \frac{W_s}{W_{real}} = \frac{h_{2s} - h_1}{h_{2r} - h_1}$$

Where,

- h_1 Is the specific enthalpy of the gas at the entrance.
- h_{2r} Is the specific enthalpy of the gas at the exit for real process.
- h_{2s} Is the specific enthalpy of the gas at the exit for isentropic process.

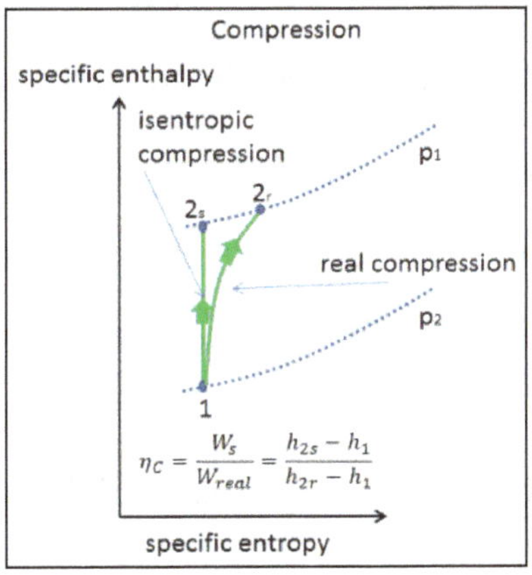

Isentropic process is a special case of adiabatic processes. It is a reversible adiabatic process. An isentropic process can also be called a constant entropy process.

Isentropic Turbine Efficiency

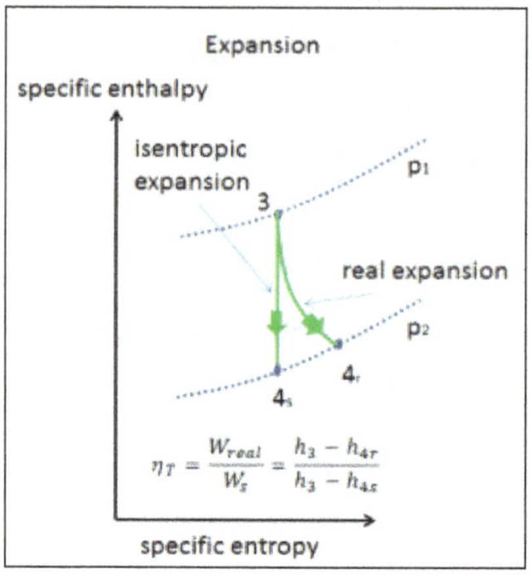

Isentropic process is a special case of adiabatic processes. It is a reversible adiabatic process. An isentropic process can also be called a constant entropy process.

Assume an isentropic expansion of helium (3 → 4) in a gas turbine. In this turbines the high-pressure stage receives gas (point 3 at the figure; p_3 = 6.7 MPa; T_3 = 1190 K (917 °C)) from a heat exchanger and exhaust it to another heat exchanger, where the outlet pressure is p_4 = 2.78 MPa (point 4). The temperature (for isentropic process) of the gas at the exit of the turbine is T_{4s}= 839 K (566 °C).

Calculate the work done by this turbine and calculate the real temperature at the exit of the turbine, when the isentropic turbine efficiency is η_T = 0.91 (91%).

Solution:

From the first law of thermodynamics, the work done by turbine in an isentropic process can be calculated from:

$$W_T = h_3 - h_{4s} \rightarrow W_{Ts} = c_p (T_3 - T_{4s})$$

From Ideal Gas Law we know, that the molar specific heat of a monatomic ideal gas is:

$$C_v = 3/2R = 12.5\ J/mol\ K \text{ and } C_p = C_v + R = 5/2R = 20.8\ J/mol\ K$$

From Ideal Gas Law we know, that the molar specific heat of a monatomic ideal gas is:

$$C_v = 3/2R = 12.5\ J/mol\ K \text{ and } C_p = C_v + R = 5/2R = 20.8\ J/mol\ K$$

We transfer the specific heat capacities into units of J/kg K via:

$$c_p = C_p \,.\, 1/M \text{ (molar weight of helium)} = 20.8 \times 4.10^{-3} = 5200\ J/kg\ K$$

Specific Heat Calculation

Using this temperature and the isentropic compressor efficiency we can calculate the heat added by the heat exchanger:

$$Q_{add} = c_p(T_3 - T_1) - (c_p(T_{2s} - T_1)/\eta_K) = 5200.(1190 - 299) - 5200.(424\text{-}299)/0.87 =$$
$$4.633\ MJ/kg - 0.747\ MJ/kg = 3.886\ MJ/kg$$

The work done on the gas by the compressor in the isentropic compression process is:

$$W_{C,s} = c_p (T_{2s} - T_1) = 5200 \times (424 - 299) = 0.650\ MJ/kg$$

The real work done on the gas by the compressor in the adiabatic compression is then:

$$W_{C,real} = c_p (T_{2s} - T_1).\ \eta_C = 5200 \times (424 - 299) / 0.87 = 0.747\ MJ/kg$$

The turbine outlet temperature of the gas, $T_{4,is}$, can be calculated using the same p, V, T relation but between states 3 and 4:

$$\frac{p_1}{p_2} = \left[\frac{V_2}{V_1}\right]^k = \left[\frac{T_1}{T_2}\right]^{\frac{k}{k-1}}$$

From the previous equation follows that the outlet temperature of the gas, $T_{4,is}$, is:

$$T_{4,is} = \left[\frac{p_4}{p_3}\right]^{\frac{k-1}{k}} .T_3 = \left[\frac{2.78}{6.7}\right]^{\frac{0.66}{1.66}} .1190 = 839\left(566^{\circ} C\right)$$

The work done by gas turbine in the isentropic expansion is then:

$$W_{T,s} = c_p (T_3 - T_{4s}) = 5200 \times (1190 - 839) = 1.825\ MJ/kg$$

The real work done by gas turbine in the adiabatic expansion is then:

$$W_{T,real} = c_p (T_3 - T_{4s}).\ \eta_T = 5200 \times (1190 - 839) \times 0.91 = 1.661\ MJ/kg$$

the thermal efficiency of an ideal Brayton cycle is a function of pressure ratio and κ:

$$\eta_{th} = 1 - \frac{1}{PR^{k-1/k}}$$

therefore,

$$\eta_{th} = 0.295 = 29.5\%$$

The thermal efficiency can be also calculated using the work and the heat (without η_K):

$$\eta_{th,s} = (W_{T,s} - W_{C,s}) / Q_{add,s} = (1.825 - 0.650) / 3.983 = 0.295 = 29.5\%$$

Finally, the thermal efficiency including isentropic turbine/compressor efficiency is:

$$\eta_{th,real} = (W_{T,real} - W_{C,real}) / Q_{add} = (1.661 - 0.747) / 3.886 = 0.235 = 23.5\%$$

References

- Wong 'Thermodynamics for Engineers',2nd Ed.,2012, CRC Press, Taylor & Francis, Boca Raton, London, New York. (ISBN 978-1-4398-4559-2)
- Carnot-Cycle, Thermodynamic-Cycles, Thermodynamics, Supplemental-Modules-(Physical-and-Theoretical-Chemistry), Physical-and-Theoretical-Chemistry-Textbook-Maps, Bookshelves: chem.libretexts.org, Re-trieved 23 March, 2019

- "Geothermal Technologies Program: Hydrothermal Power Systems". Geothermal Technologies Program: Technologies. U.S. DOE Energy Efficiency and Renewable Energy (EERE). 2010-07-06. Retrieved 2010-11-02
- Brayton-cycle-gas-turbine-engine, thermodynamic-cycles, nuclear-engineering: nuclear-power.net, Retrieved 24 April, 2019
- Ronald dipippo (2007). Geothermal Power Plants, Second Edition: Principles, Applications, Case Studies and Environmental Impact. Oxford: Butterworth-Heinemann. P. 159. ISBN 0-7506-8620-0

3

Steam Generators

A power generation device which is used to generate steam by applying heat energy to water is referred to as steam generator. They can be categorized into water tube boilers and fire tube boilers. This chapter has been carefully written to provide an easy understanding of these classifications of steam generators as well as their working.

A steam boiler is a type of closed container, designed with steel for heating the water to generate steam by some energy source like burning of fuel with eventually. The vapor generated may be delivered at low-pressure for the industrial progression work in sugar industries, cotton mills, and for generating steaming water that can be utilized for heat up fixing at much low-force. The capacity of the boiler should have ten liters of water and the working pressure must be 3.4 Kgf/cm^2 (kilogram-force).

What is a Steam Boiler?

A steam boiler is a power generation device, used for generating steam by applying the heat energy to water. The pressure range of earlier boilers ranges from low pressure to medium pressure (7 kPa to 2000 kPa/ 1psi to 290 psi). The present boilers are more useful because it works with high pressure than old ones. This boiler is very used whenever a steam source is necessary, and the size, type mainly depends on the type of application like mobile steam engines which includes handy engines, steam locomotives, and road vehicles. These vehicles include a mini boiler that can be worked with steam power. Generally, power stations or stationary steam engines have a separate large steam generating capacity.

Steam Boiler.

The Function of Steam Boiler

The main function of a steam boiler is to producing, storing and troubling the vapor. The liquid contained boiler is nothing but a shell and the heat energy produced while burning of fuel will be moved to water, and then it converts into steam at the required pressure as well as temperature.

The main conditions of this boiler mainly include the water container should lock very carefully. The vapor of water should be supplied in the preferred conditions namely, quality, rate, pressure, and temperature.

Steam Boiler Working Principle

The main working principle of a steam boiler is easy. This boiler is one type of closed device in the shape of a cylindrical. The capacity of the boiler to have steam, as well as water is sufficient.

Generally, the liquids are stored in the boiler for generating steam by burning the fuels or applying the heat energy at different conditions of pressure based on the size of the vessel as well as specifications. Finally, the steam in the boiler supplies using a pipe and flows into various industries like plants.

The main components of this boiler mainly include a shell, furnace, grate, mountings, water space, accessories, refectory, level of water, scale, foaming, lagging, and blowing off.

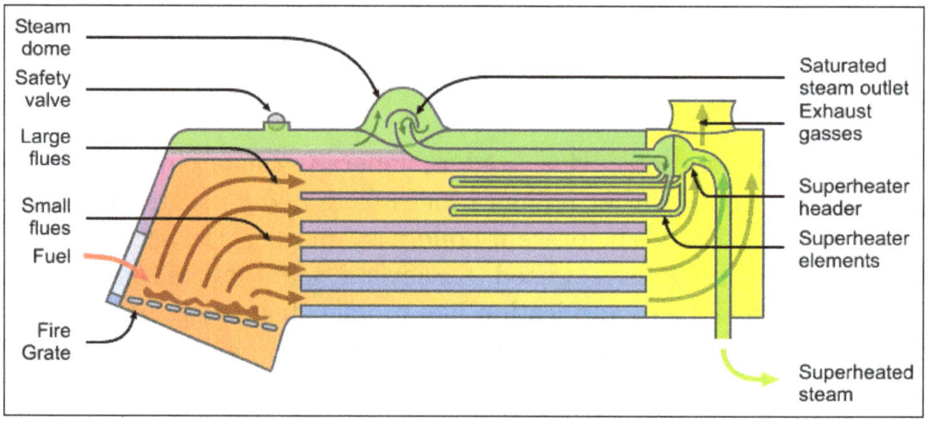

Fire Tube Boiler.

Advantages of Steam Boiler

The advantages of this boiler include the following:

- The construction cost of these boilers is low.
- This boiler uses any kind of chimney.
- It occupies less floor area.
- It is portable.
- It has a self-enclosed boiler.

Disadvantages of Steam Boiler

The disadvantages of this boiler include the following:

- The design of the steam boiler is vertical so the capacity of steam rising is low.
- It has limited pressure and capacity.
- The cleaning and examination of these are difficult.
- It requires high headspace.

Applications of Steam Boiler

The applications of this boiler include the following:

- These are used to generate power in steam turbines or engines.
- These are used for various processes in process industries.
- These are used in houses or buildings in cool weather for generating a hot water supply.

Characteristics of Steam Boiler

The characteristics of steam boiler include the following:

- Steam boiler generates the highest amount of steam with low fuel utilization.
- It must be less weight as well as needs small space.
- It must be immediate starting.
- These boilers must be cheap as well as needless concentration.
- These boilers handle any kind of fluctuating load.

Classification of Boilers

A gas/oil central heating boiler (heat generator) is like the engine of a car, this provides the heat that the facility needs to warm itself up. The size of the boiler is matched to the size of the facility.

- If the boiler is oversized, the fuel bills will be excessive.
- If the boiler is undersized, it may not generate enough heat in winter.

The ideal size for a boiler is one that just copes adequately on the coldest day of the year. Most boilers are oversized by at least 30%. This is due to the way systems used to be calculated with a card calculator. These were always over-calculated "to be on the safe side." Today, the emphasis is on energy conservation, and the fact that heat loss calculations can be done very accurately, means there is no need to oversize. This allows smaller radiators and less water in the system, which in turn, means a smaller boiler and reduced costs for both installation and fuel bills.

The boiler does not directly govern the amount of radiators fitted to the system. It is the power of the pump and circulation of the water through adequately sized pipes that determines the number of radiators you can have. But the total output of all the radiators, pipes, and cylinders determines the size of the boiler.

The boiler is not the heating system; it is only one of the parts in the global heating system. A heating system consists of four main parts:

1. Boiler/burner combination (the part producing the heat).
2. Piping with pumps and valves (the part distributing the heat).
3. Radiators and convectors (the part emitting the heat to the room).
4. Control equipment such as room thermostat and outside temperature control (the part controlling room and water temperature).

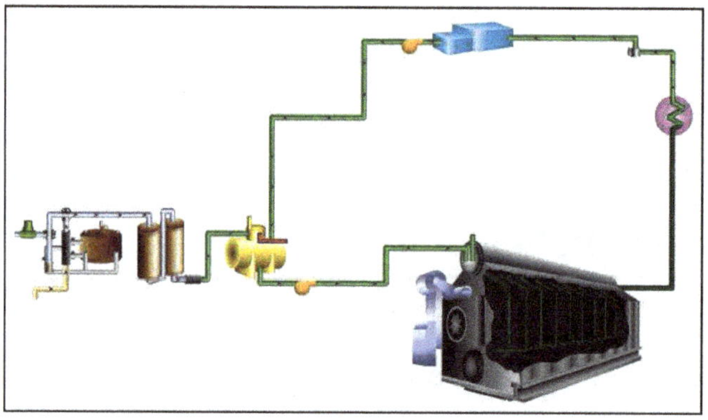

Different Parts of a Heating System.

Boiler Types and Classifications

There are two general types of boilers: fire-tube and water-tube. Boilers are classified as high-pressure or low-pressure and steam boiler or hot water boiler. Boilers that operate higher than 15 psig are called high-pressure boilers.

A hot water boiler, strictly speaking, is not a boiler. It is a fuel-fired hot water heater. Because of its similarities in many ways to a steam boiler, the term hot water boiler is used.

- Hotwater boilers that have temperatures above 250 °Fahrenheit or pressures higher than 160 psig are called high temperature hot water boilers.
- Hotwater boilers that have temperatures not exceeding 250 °Fahrenheit or pressures not exceeding 160 psig are called low temperature hot water boilers.

Heating boilers are also classified as to the method of manufacture, i.e., by casting (cast iron boilers) or fabrication (steel boilers). Those that are cast usually use iron, bronze, or brass in their construction. Those that are fabricated use steel, copper, or brass, with steel being the most common material.

Steel Boilers

Steel boilers are generally divided into two types: firetube and water-tube.

Fire-tube Boilers

In fire-tube boilers, combustion gases pass through the inside of the tubes with water surrounding the outside of the tubes. The advantages of a fire-tube boiler are its simple construction and less rigid water treatment requirements.

The disadvantages are the excessive weight-per-pound of steam generated, excessive time required to raise steam pressure because of the relatively large volume of water, and inability to respond quickly to load changes, again, due to the large water volume.

The most common fire-tube boilers used in facility heating applications are often referred to as scotch or scotch marine boilers, as this boiler type was commonly used for marine service because of its compact size (fire-box integral with boiler section).

The name "fire-tube" is very descriptive. The fire, or hot flue gases from the burner, is channeled through tubes that are surrounded by the fluid to be heated. The body of the boiler is the pressure vessel and contains the fluid. In most cases, this fluid is water that will be circulated for heating purposes or converted to steam for process use.

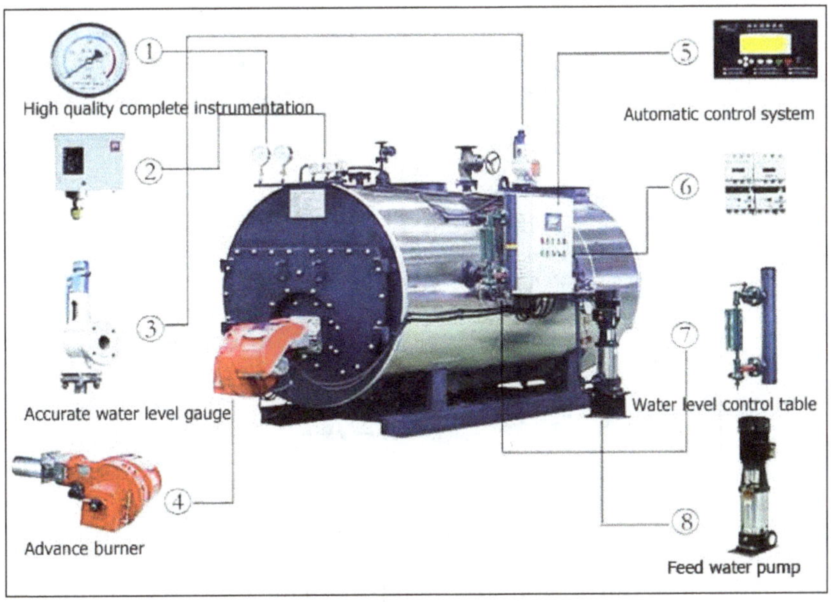

Fire-tube Boiler Gas Flow.

Every set of tubes that the flue gas travels through, before it makes a turn, is considered a "pass". So, a three-pass boiler will have three sets of tubes with the stack outlet located on the rear of the boiler. A four-pass boiler will have four sets and the stack outlet at the front.

Advantages of fire-tube boilers are:

- Relatively inexpensive.
- Easy to clean.
- Compact in size.
- Available in sizes from 600,000 btu/hr to 50,000,000 btu/hr.

- Easy to replace tubes.
- Well suited for space heating and industrial process applications.

Disadvantages of fire-tube boilers include:

- Not suitable for high pressure applications 250 psig and above.
- Limitation for high capacity steam generation.

Water-tube Boilers

In a water-tube boiler, the water is inside the tubes and combustion gases pass around the outside of the tubes. The advantages of a water-tube boiler are a lower unit weight-per-pound of steam generated, less time required to raise steam pressure, a greater flexibility for responding to load changes, and a greater ability to operate at high rates of steam generation.

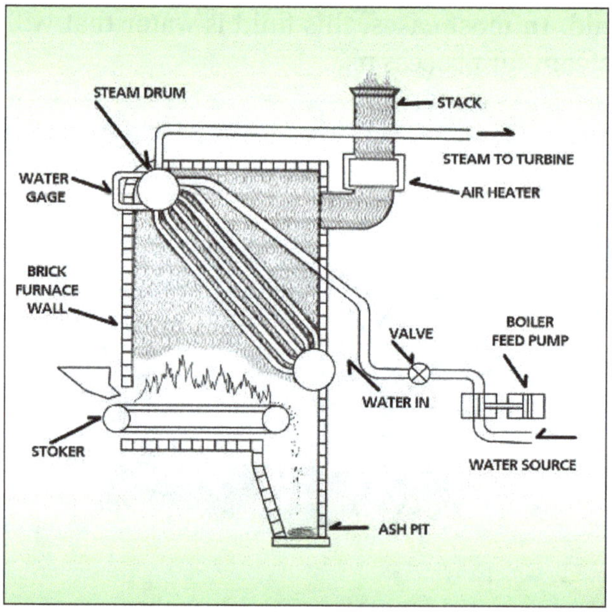

Water-tube Boiler.

A water-tube design is the exact opposite of a fire-tube. Here, the water flows through the tubes and is encased in a furnace in which the burner fires. These tubes are connected to a steam drum and a mud drum. The water is heated and steam is produced in the upper drum.

Large steam users are better suited for the water-tube design. The industrial water-tube boiler typically produces steam or hot water primarily for industrial process applications, and is used less frequently for heating applications. The best gauge of which design to consider can be found in the duty in which the boiler is to perform.

Advantages of Water-tube boilers:

- They are available in sizes far greater than a fire-tube design, up to several million pounds-per-hour of steam.
- They are able to handle higher pressures up to 5,000 psig.

- They recover faster than their fire-tube cousin.
- They have the ability to reach very high temperatures.

Disadvantages of the water-tube design include:

- High initial capital cost.
- Cleaning is more difficult due to the design.
- No commonality between tubes.
- Physical size may be an issue.

Package Boiler

A packaged boiler is a separate application which is used to get the energy from an incorporated burner with fueled by oil or gas. This type of boiler is proficient in producing huge steam output at a high-pressure as well as high-temperature. This boiler operates variable loads which have a small start-up or response time. The applications of this boiler mainly include process steam for industrial, chemical or used as power generator with a steam turbine. The package boiler can be used as a peak-load boiler because of an addition to another power that kicks in while other supplies not succeed.

Packaged Boiler.

Stoker Fired Boiler

The stoker fired boiler is most capable in common range and cogeneration systems in sugar factories. These boilers have membrane design and operate completely automatic. These Boilers have intrinsic features for the operation of trouble-free and high-thermal efficiency. These are classified based on the technique of supply fuel to the boiler & with the grate type. The stoker-fired boilers are classified into two types namely chain-gate or traveling grate stoker and spreader stoker.

Cast Iron Boilers

Cast iron boilers are made in three general types: horizontal-sectional, vertical-sectional, and one-piece. Most of the sectional boilers are assembled with push nipples or grommet type seals, but some are assembled with external headers and screw nipples. Horizontal-sectional, cast iron boilers are made up of sections stacked one above the other, like pancakes, and assembled with push nipples. Vertical-sectional, cast iron boilers are made up of sections standing vertically, like slices in a loaf of bread. One-piece cast iron boilers are those in which the pressure vessel is made as a single casting.

Cast Iron Boiler.

Steam and Condensate Boiler System

Boilers are generally used to provide a source of steam or hot water for facility heating and process needs.

In steam and condensate systems, heat is added to water in a boiler causing the water to boil and form steam. The steam is piped to points requiring heat, and as the heat is transferred from the steam to the building area or process requiring heat, the steam condenses to form condensate. In some very low-pressure saturated steam heating applications, the steam distribution piping may be sized to slope back to the boiler so that the steam distribution piping also acts as the condensate return piping (single-pipe system).

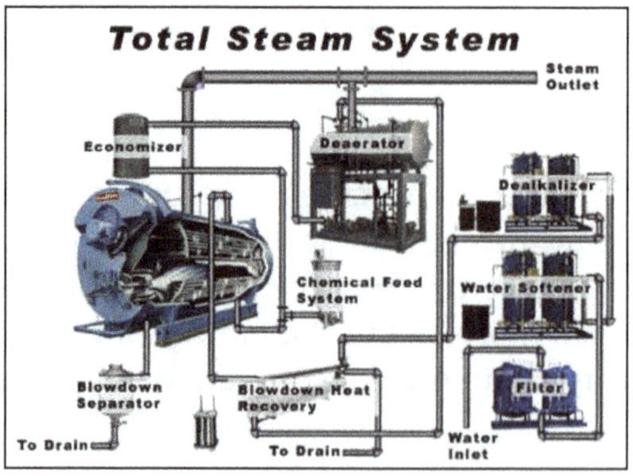

Steam and Condensate Boiler System.

In other low-pressure applications, there may be steam supply piping and condensate return piping (two-pipe system), although the condensate system is open to the steam system. In typical packaged steam boiler operations, the boiler system may generate steam at about 150 psig for distribution throughout the facility and may be lowered to the operating pressure of equipment supplied through point-of-use pressure reducing stations. As heat is transferred from the steam, condensate is formed which collects in discharge legs until enough condensate is present to operate a trap that isolates the steam distribution system from the condensate system. In common facility heating applications, the condensate system is at atmospheric pressure and the system is arranged to drain the condensate to a central condensate receiver, or into local smaller receivers that pump the condensate back to the central condensate receiver.

Hydronic Boiler System

A boiler is used to heat water that is circulated through a closed loop piping system for general facility and service water heating. Low-temperature systems generally operate below 200° Fahrenheit Medium-temperature systems generally operate at temperatures between 200° and 250° Fahrenheit.

A feature of hot water systems is an expansion tank to accommodate the expansion of the water in the system as the water is heated. The expansion tank, when piped into the system on the suction side of the circulating pumps, also pressurizes the system to prevent flashing in the circulating pump, piping, and piping components. In many low- and medium-pressure systems, pressurization is maintained by flash steam in the expansion tank. In a few hot water systems, pressurization is maintained by maintaining a compressed gas blanket above the water level in the expansion tank.

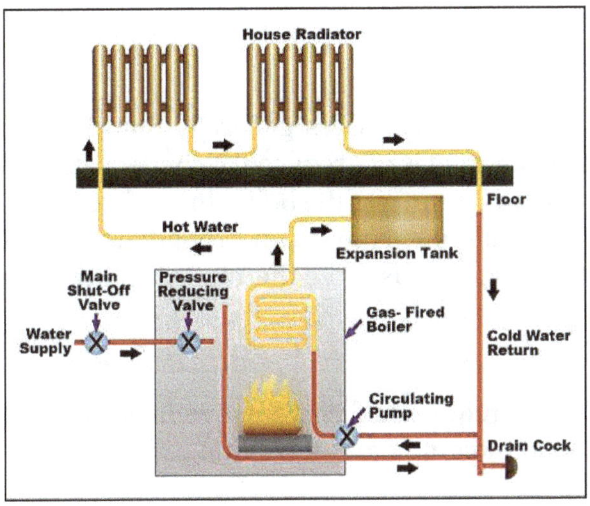

Hydronic (Hot Water) Boiler System.

High-temperature hot water systems, which operate above 250 °Fahrenheit, are basically the same as hot water systems that operate below 250 °F. High-temperature systems are generally installed when a process requires the higher temperature, a number of locations require small quantities of low-pressure steam that the high-temperature hot water can generate in a local converter, or high-temperature drop equipment can be used at end use points to minimize the size of water circulation piping required.

Most facility boiler systems are fired using a combustible gas (typically natural gas or propane) or fuel oil. In many facilities, the boilers are designed to fire both a combustible gas fuel and a fuel oil. In these facilities, the combustible gas fuel is generally natural gas that is considered the primary fuel, and fuel oil is considered to be the backup fuel.

Boiler System Major Components

Boiler systems are comprised of the major components described below and shown in figure.

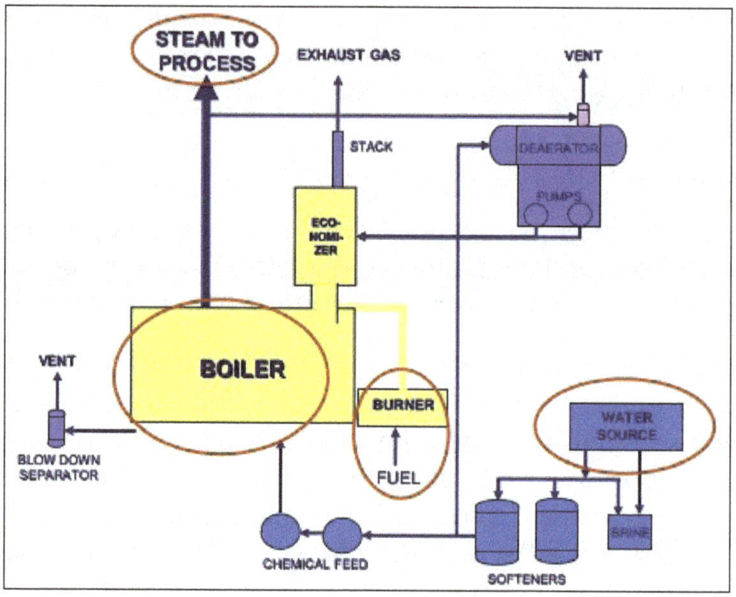

Boiler Room Schematic.

Feedwater Heaters

Feedwater heaters are energy recovery devices generally found only in large steam generating plants where all of the steam generated is not reduced to condensate by the steam user. This "waste steam" is reduced to condensate for return to the boiler in the feedwater heater. The boiler feedwater is used as a cooling medium to reduce the steam to condensate, which increases the temperature of the feedwater and, thereby, increases the thermal efficiency of the boiler.

Fuel Heater

Many boilers firing heavy fuel oil require fuel heaters to reduce the fuel viscosity, so the fuel can be atomized by the burner system for complete combustion.

Deaerators

A deaerator is a special case of feedwater heater that is designed to promote the removal of non-condensable gases from the boiler feedwater. The principal gases of concern are oxygen, carbon dioxide, and ammonia, which are major contributors to boilers, and steam and condensate piping corrosion problems. In small steam plants, a portion of the steam generated by the boiler is used to operate the deaerator if "waste steam" is not available. Failure to maintain and properly operate the deaerator can lead to early failure of the boiler, steam using equipment, and the steam and condensate piping.

Pumps

In most hot water systems, the system circulating pumps are electric motor-driven, end suction centrifugal pumps. In steam systems, the condensate return pumps are typically electric motor-driven, end suction, centrifugal or turbine-type pumps. Feedwater pumps are generally elec-tric motor-driven, multiple-stage, end suction centrifugal pumps. The shutoff head of the pump must be greater than the steam or hot water system operating pressure.

Combustion Air Blowers

In many packaged boiler installations, the combustion air fan is designed and provided by the boiler manufacturer and is integral with the boiler housing. In installations where a stand-alone fan is provided, low-pressure centrifugal blowers are commonly used. An important characteristic of the blower is the ability to maintain a relatively constant air pressure over a wide range of airflows.

Flue

Flues (boiler firebox exhaust duct or boiler discharge stack) must be large enough to conduct the products of combustion away from the boiler with a minimum of duct friction loss. Flues may be fabricated from any material suitable for the operating temperature and pressure. Common materials of construction associated with packaged boiler installations are carbon steel and stainless steel.

Economizer

An economizer is an energy recovery device that uses the hot exhaust gases from the boiler (waste heat) to heat combustion air or feedwater.

Steam Traps

Steam traps are installed throughout steam systems to remove condensate (spent steam), air, and non-condensable gases from the steam system. There are five types of steam traps in general use today, as described below:

1. The heart of a balanced pressure thermostatic trap is the flexible bellows that moves the valve head from its seat to discharge the condensate. The bellows is filled with a volatile fluid and hermetically sealed. The fluid has a pressure-temperature relationship that closely parallels, but is approximately 10 degrees Fahrenheit below that of steam.
2. The liquid expansion steam trap has for its operating element a liquid-filled cartridge. Within this cartridge is a hermetically sealed bellows which is attached to the valve head and plunger.
3. Float and thermostatic steam traps provide immediate and continuous discharge of condensate, air, and non-condensables from a steam system as soon as they reach the trap. The trap consists of a ball float connected by a lever assembly to the main valve head. As condensate reaches the trap, the ball float rises, positioning the valve to discharge the condensate at the same rate as it reaches the trap.
4. The inverted bucket steam trap is a type of trap with an inverted bucket attached to the valve head by a lever mechanism and operates to open and close the trap. When condensate enters the trap, a water seal is formed around the bottom of the inverted bucket which, since it is filled with air, becomes buoyant and rises and closes the trap. A small hole in the top of the inverted bucket allows air to escape with condensate taking its place inside the

bucket. The inverted bucket loses its buoyancy and sinks to the trap bottom, opening the valve to discharge the condensate.

5. Thermodynamic steam traps are a type of steam trap that responds to differences in kinetic energy between steam and condensate to open and close the valve for discharging condensate.

Piping

Piping two-inches and smaller used in steam and hot water systems is typically Schedule 80, American Society for Testing and Materials (ASTM) A 106, Standard Specification for Seamless Carbon Steel Pipe for High-Temperature Service, Grade B, steel pipe with threaded joints and carbon steel fittings. Piping larger than two-inches is typically standard weight, ASTM A 106, Grade B, steel pipe with flanged joints and carbon steel fittings.

Boiler Room Definition/Terminology

There are many terms used in a discussion of boilers, the following is a list of some of the most common terms. There is a glossary provided that covers some of the other terms that may be also used. This topic also contains some of the basic valves that are used on boiler and boiler systems, along with some of the common HVAC and piping symbols.

HVAC and Boiler Symbols

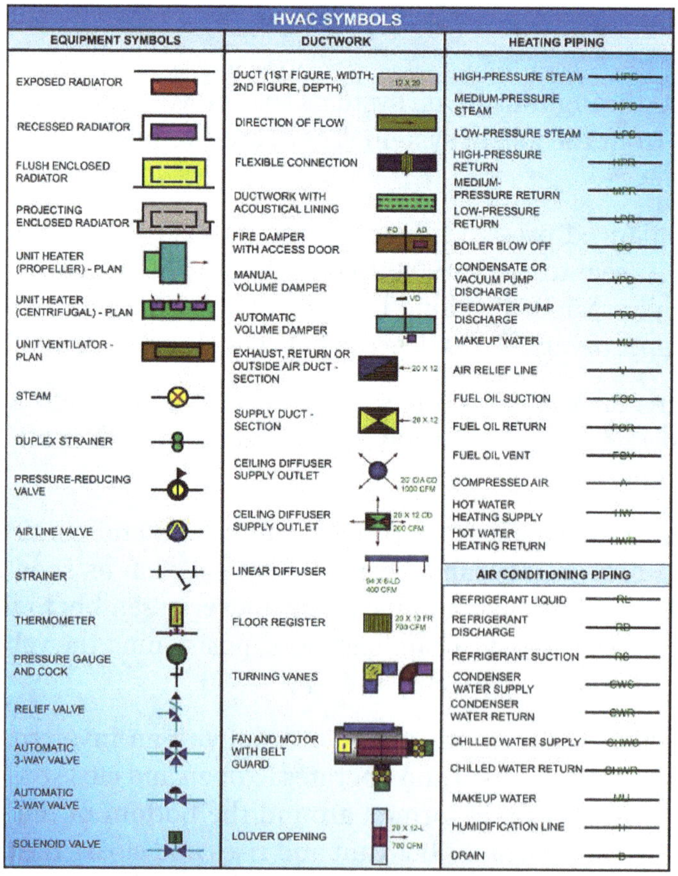

Common HVAC and Boiler Symbols.

Basic Boiler

The following valves are the most common valves found on boilers and boiler systems. Each type is briefly discussed.

Butterfly Valves

A butterfly valve is composed of two semicircular plates hinged on a common spindle, used to permit flow in one direction only. Butterfly valves are of the quarter-turn family and are so designed because a 90-degree turn of the operator fully opens or closes the valve.

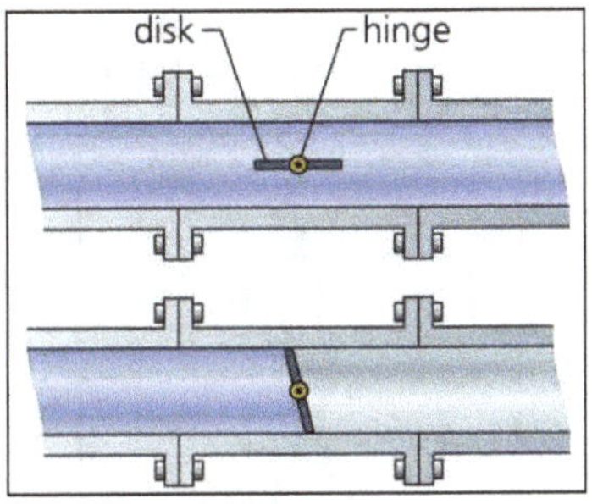

Butterfly Valve.

The valve uses elastomer seats and seals and their surge in popularity can be attributed to these advantages. In some cases, they may be used for non-critical throttling applications. They are lighter in weight than conventional valves. The position of the lever indicates whether they are wide open, partially open, or fully closed.

Butterfly valves are compact and space-saving and easily installed in new piping or as replacements in existing piping. They are easily adapted to lever, manual, gear, electric, or pneumatic operation.

Gate Valves

Gate valves are, by far, the most widely used in industrial piping. They are used as stop valves to fully shut off or fully turn on flow – the only job for which gate valves are recommended.

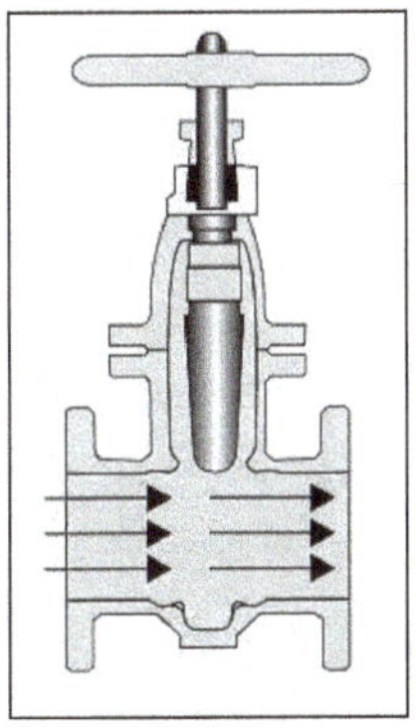

Gate Valve

Gate valves are inherently suited for full open service. Flow moves in a straight line and practically without resistance when the wedge is fully raised. There are two basic designs of gate valves: inside screw stem and outside screw stem.

Seating is perpendicular or at right angles to the line of flow – meets it head on. That is one reason why gate valves are impractical for throttling service. When throttling is necessary, globe valves should be used. Repeated movement of the wedge near the point of closure, under high velocity flow, may create a drag on the seating surfaces and cause galling or scoring on the downstream side. A slightly opened wedge may also cause turbulent flow with vibration and chattering of the wedge.

A gate valve usually requires more turns to open it fully. Also, unlike many globe valves, the volume of flow through the valve is not in direct relation to the number of turns of the hand-wheel.

Globe Valves

Unlike the perpendicular seating in gate valves, globe valve seating is parallel to the line of flow.

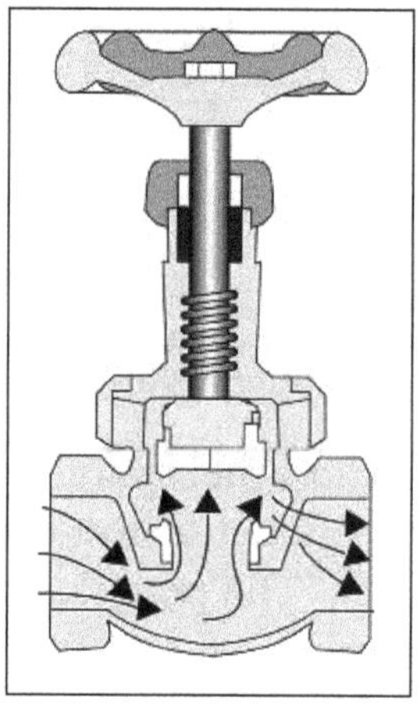

Globe Valve.

All contact between seat and disc ends when flow begins. These are advantages for more efficient throttling of flow, with minimum wire drawing and seat erosion. Valve disks and seats in most globe valves can be conventionally repaired or replaced – often without removing the valve body from the line.

Shorter disk travel, with fewer turns required to operate globe valves, saves considerable time and work, also wear on valve parts.

Check Valves

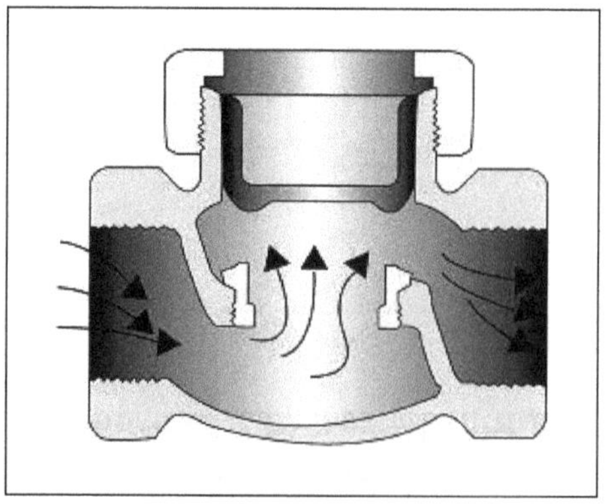

Check Valve

Check valves are designed to automatically prevent the reversal of flow in a pipeline system. They control the direction of flow, rather than throttling or isolating flow as other valve designs do. Reverse flow may create problems or it could cause damage to equipment. Check valves are sometimes known as "reflux" valves.

There are several basic designs of check valves, a few of which are described below.

Swing Check Valves

This type of check valve uses a hinged mounted disc that swings open and closed with flow. They can be used in the horizontal and vertical (flow upwards) position. The swing check is the most commonly used design of check valve as it does not restrict flow.

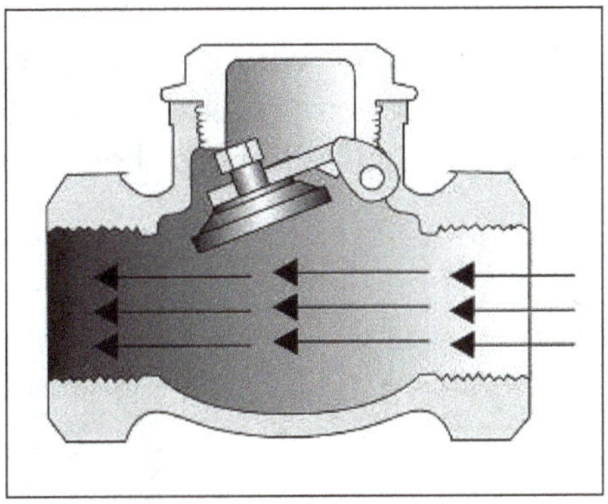

Swing Check Valve.

Lift or Piston Check Valve

This type of check valve uses a piston rather than a hinge-mounted disc to prevent the reversal of flow. This provides a cushioning effect during the operation of the valve. They must only be used in a horizontal position. Lift check valves, like globe valves, are flow restricting; therefore, they are generally used as companions to globe valves.

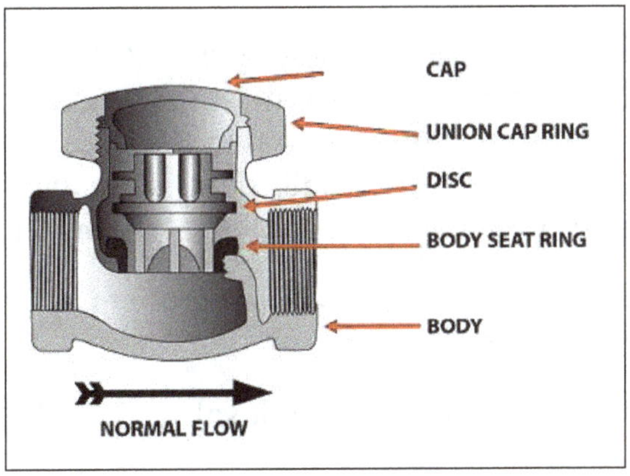

Lift Check Valve.

Pressure Relief Valves

Pressure relief is simply a dumping of excess fluid safely into the atmosphere. The excess fluid is that which would cause pressure to exceed the safety limit. The relief/safety valve is the most widely used piece of equipment in this category. However, liquid seals and rupture discs may also be used.

There are two basic kinds of relief valves: self-operated and pilot-operated. The spring-type relief valve is the most widely used. The pilot-operated type is also frequently used, and it offers more precise operation. The pilot-operated type is more frequently used as pressures become higher and capacities greater.

Pressure relief valves are designed to open automatically at a pre-determined set pressure level of system pressure and to achieve a rated relieving capacity at a specified pressure and temperature above the setpoint (overpressure) before re-closing at a pressure below the opening point (blowdown). The simplest and most reliable type of pressure relief valve, even some four-hundred years on from the first design, is the spring-loaded design where a spring force opposes the system pressure acting on the valve disc. When the system pressure rises above the level of the spring force, the valve opens. This valve type may also be fitted with a bellows for better emission control performance.

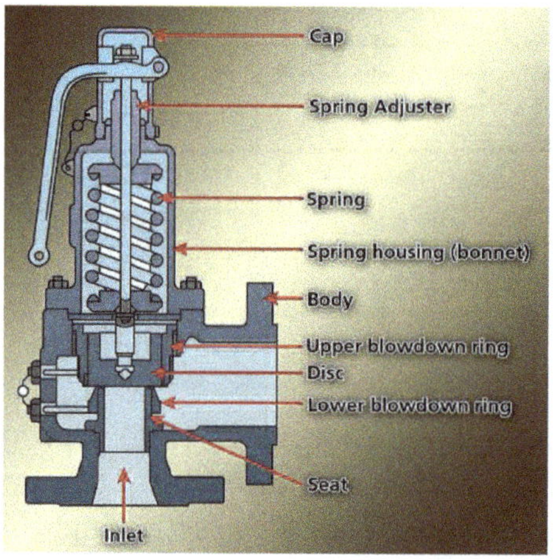

Spring-Type Pressure Relief Valve.

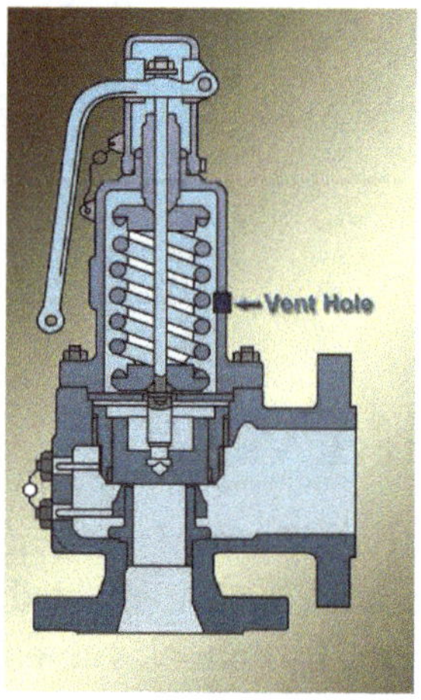

Bellow Spring-Type Relief Valve.

The significant elements of all spring-loaded pressure relief designs are the springs and the seats. The springs must provide the desired compression rate and a reasonable range of adjustment. They must also fit into the valve bonnet and stay within the design perimeters. The seats may be flat or angled, metal or soft. As the seat area usually defines the load transmitted to and from the spring, very high precision is essential to ensure proper valve operation.

A second type of valve, which is more sophisticated and offers operating advantages in selected applications, is the pilot-operated pressure relief valve.

This type of valve consists of a main valve and a pilot valve. The pilot responds directly to system pressure and communicates with the main valve. As with the spring-loaded valve, many unique models exist. However, some common design features of pilot-operated pressure relief valves include the sensing line, pilot valve, and main valve.

The sensing line is either connected to the valve inlet or a remote location and conveys system pressure to the pilot.

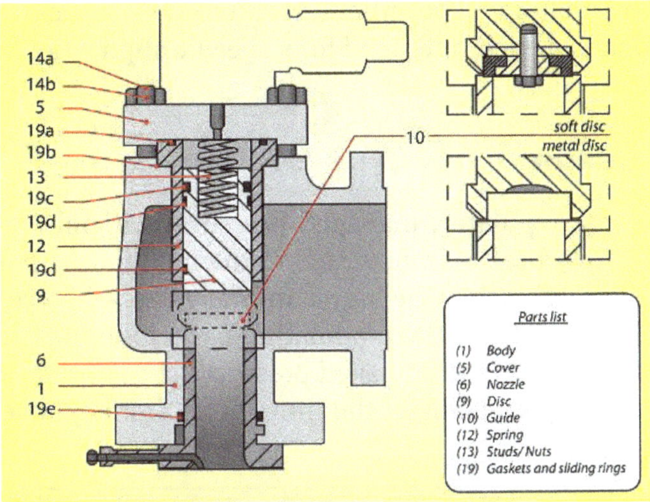

Pilot-Operated Relief Valve.

The pilot valves senses and responds to the system pressure. The pilot is the controlling member of the valve system and determines all of the operating characteristics of the valve. It consists of many small parts and passages and usually relies on elastomer seals for operation.

The main valve operates in response to the pilot and provides the main rated flow capacity to reduce excess pressure.

Regulating Valves

Regulating valves are used in many systems on a boiler from feed flow to fuel flow. Regulators attempt to manage flow by adjusting the flow area available to the fluid. Most designs incorporate a plug or similar element that occupies a portion of a stationary orifice while throttling. In an action often called "modulation", the plug or similar element is shifted to increase or decrease available flow area. Flow through a particular size of open orifice area is primarily dependent upon the available pressure difference.

Where the managed parameter is pressure, obviously, flow and pressure are interdependent. While the pressure being maintained may be at a location remote from the valve, there must always be an immediate and inherent relationship maintained among the managed flow, the pressure to which the regulator is responding and the regulator's response mode. Essentially, the regulator must be able to influence pressure at the measured point and must exercise that influence in the appropriate direction of response.

Where the managed pressure is upstream from the regulating valve, you normally would want the valve to open to permit flow as pressure begins to rise.

Therefore, the valve's response disposition is "open on rise." The term normally applied to this mode of operation is inlet pressure regulator function. Conversely, if managed pressure is downstream from the valve, the valve's disposition would be "open on drop," and the normal reference for function is outlet pressure regulator.

Finally, regulators often are called upon to manage operating conditions other than pressure, such as temperature or flow. A regulator will be either direct- or pilot-operated. Direct-operated regulators are simple, single-minded devices intended for a specific application within a particular range of pressures.

Direct-operated Regulator

In a direct-operated pressure regulator, managed pressure is applied to some internal surface within the valve. The resulting force is transmitted to the modulating parts of the valve and balanced by a second force applied within the valve. In many cases, the second force is adjustable to establish the setpoint. Movement of the modulating parts inherently involves a change in the balanced forces and thus a change in the managed pressure. The more the demanded flow deviates from what the regulator was experiencing at the time it was first adjusted, the more the pressure must deviate from setpoint.

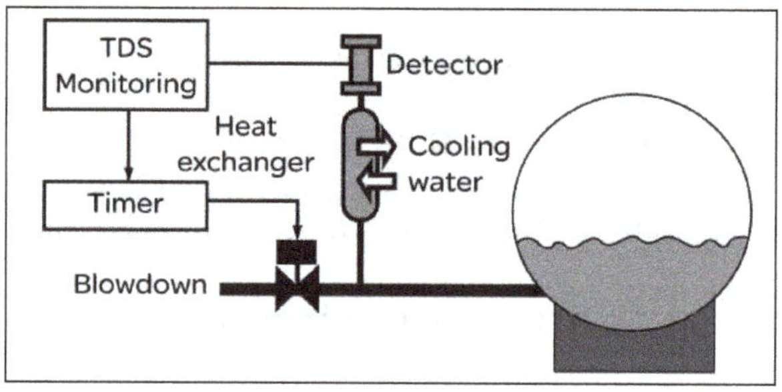

Blowdown in boiler.

Pilot-operated Regulator

A pilot-operated regulator consists of a pilot section and a slave section. Two flows are managed by the valve. The first, which is by far the greater proportion of the total flow, is managed by the plug and orifice in the slave section. The second, the pilot stream, is managed by the pilot section. The pilot section may have its inlet connected to the immediate upstream side of the main valve or may source its flow from a higher pressure space. The pilot section's modulating portion is available in all dispositions and functions of direct-operating regulators.

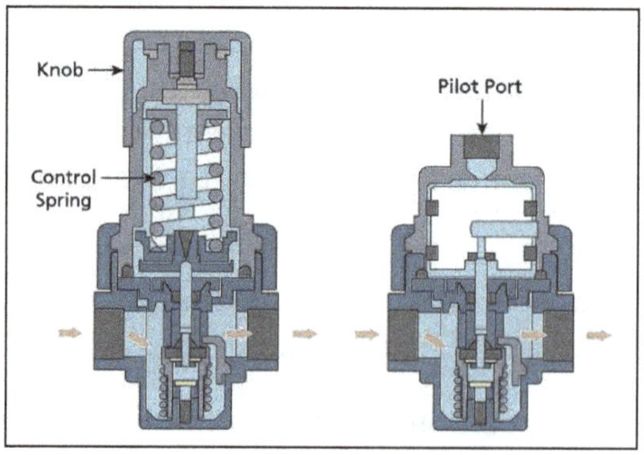

Pilot-Operated Regulator.

Boiler Blowdown

Blowdown of steam boilers is very often a highly neglected or abused aspect of routine boiler room maintenance. The purpose of boiler blowdown is to control solids in the boiler water. Blowdown protects boiler surfaces from severe scaling or corrosion problems that can result otherwise.

There are two types of boiler blowdowns: continuous and manual.

A continuous blowdown uses a calibrated valve and a blowdown tap near the boiler water surface. As the name implies, it continuously takes water from the top of the boiler at a predetermined rate. A continuous blowdown is an optional feature and may not be included on your steam boiler; however, all steam boilers should include a means for manual blowdown as standard equipment.

Manual blowdowns are accomplished through tapings at the bottom of the boiler. These openings allow for the removal of solids that settle at the bottom of the boiler. Manual blowdown is also used to keep water level control devices and cutoffs clean of any solids that would interfere with their operation. All steam boilers require manual blowdown whether or not they are supplied with continuous blowdowns.

Frequency of Manual Blowdown

When continuous blowdown is used, manual blowdown is primarily used to remove suspended solids or sludge. The continuous blowdown removes sediment and oil from the surface of the water along with a prescribed amount of dissolved solids.

When surface or continuous blowdown is not used, manual blowdown is used to control the dissolved or suspended solids in addition to the sludge.

In practice, the valves of the bottom blowdown are opened periodically in accordance with an operating schedule and/or chemical control tests. From the standpoint of control, economy, and results, frequent short blows are preferred to infrequent lengthy blows. The length and frequency of the blowdown is particularly important when the suspended solids content of the water is high. With the use of frequent short blows, a more uniform concentration of the pressure vessel water is maintained.

In cases where the feedwater is exceptionally pure, or where there is a high percentage of return condensate, blowdown may be employed less frequently since less sludge accumulates in the pressure vessel. When dissolved and/or suspended solids approach or exceed predetermined limits, manual blowdown to lower the concentrations is required.

It is generally recommended that a steam boiler be blown down at least once in every eight-hour period, but frequency may vary depending upon water and operating conditions. The blowdown amounts and schedule should be recommended by your local Cleaver-Brooks authorized representative.

A hot water boiler does not normally include openings for surface blowdown and bottom blowdown since blowdowns are seldom practiced. Always be alert to system water losses and corresponding amount of raw water makeup. A water meter is recommended for water makeup lines.

Proper blowdown is performed as follows:

1. Blowdown should be done with the boiler under a light load.
2. Open the blowdown valve nearest the boiler first. This should be a quick-opening valve.

3. Crack open the downstream valve until the line is warm. Then open the valve at a steady rate to drop the water level in the sight glass 1/2 inch. Then close it quickly being sure that the handwheel is backed off slightly from full close to relieve strain on the valve packing.
4. Close the valve nearest the boiler.

Repeat the above steps if the boiler has a second blowdown tapping. Water columns should be blown down at least once a shift to keep the bowls clean. Care should be taken to prevent low water shutdown if this will affect process load.

Please keep in mind that all blowdown piping should be checked once a year for obstructions.

Blow Down Purpose

The purpose of blow down is to control the amount of solids and sludge in the boiler water. The blow down process involves partially draining the boiler to remove sludge and to maintain pre-determined concentration levels of solids.

As the water is turned into steam, the solids remain behind. Unless there is 100% condensate return, the solid content tends to build up when the boiler takes on make-up water. On hot water systems, there is generally no make-up water. Therefore, the solid concentration remains constant and no blow down is needed.

The amount and frequency of blow down differs for each boiler application and should be determined by your water management consultant. Blow down is affected by the type of boiler, operating pressure, water treatment, and the amount and quality of make-up water.

Blow down piping should be at least the same size as the blow down tapping on the boiler. Blow down valves should be sized in accordance with the ASME code and piped to a safe point of discharge. There should be either two slow opening valves or one quick opening valve and one slow opening valve piped in series. A slow opening valve is defined as needing five complete 360 degree turns to go from fully closed to fully open. A quick opening valve goes from fully closed to fully open in one complete motion. In the case of one quick and one slow opening valve, the quick opening valve should be located closest to the boiler. If possible, the blow down valves should be piped on the same side of the boiler as the water column gauge glass.

To blow down the boiler:

- Open the quick opening valve (valve closest to the boiler) first.
- Open the slow opening valve last.
- Blow down the boiler for the required amount of time, per your water management consultant, by opening and then closing the slow opening valve.

Pay close attention to the water level in the gauge glass. Certain loads may require several blow down cycles of short duration to maintain proper water level in the boiler.

- Close the slow opening valve first.

- Close the quick opening valve (the valve closest to the boiler) last.
- Open the slow opening valve again to drain the line between the quick and slow opening valve.
- Close the slow opening valve again and double-check for tight shutoff after the valve has cooled off.

Never pump the quick opening valve to blow down the boiler. This may cause water hammer, which could damage piping and valves and may cause personal injury. Also, never leave an open blow down valve unattended.

The quick opening valve (the valve closest to the boiler) is opened first and closed last, which ensures its protection from the wear associated with blow down. This will make this valve more reliable so maintenance and repair can be performed on the slow opening valve furthest from the boiler, without draining the boiler.

Fire Tube Boilers

The initial steam generators, also called boilers, required a fire box for the fuel. These started out being wood-burning and quickly changed to coal-burning. The fire box has tubes running through the water chamber, heating the water to steam and then releasing the fuel fume gasses out through a smoke stack. Railroad train engines and boats were the first to use this type of steam generation for power.

Water Tube Boilers

Water tube boilers came about so that the steam generated could be produced at a higher pressure. The water flowed through tubes at an angle while the heat flowed up and around the tubes. The higher steam pressure gave greater force to push a piston or turn a turbine wheel utilizing less heating fuel.

Combustion Heat Generator

Combustion heat generators follow a similar heat exchange concept as the tube boilers, but can product even higher pressures for power. These are used mainly in power plants producing electricity. Their steam pressures can almost meet or, in some super-critical steam-designed plants, exceed the critical water pressure of 221 bar. The temperature of the steam at these highly compressed rates can reach over 500 degrees Celsius.

Heat Recovery Steam Generator

The heat recovery steam generator, or heat exchanger, recovers the high pressure hot gas steam and utilizes that steam after running it through a chain of heat exchanges to run other lower-powered machines. This recovered steam can even be used at these lower pressures to supply steam heat for other industrial buildings or even homes.

Nuclear Power Plant Steam Generators

There are two major types of nuclear steam generators; the (BWR), Boiling Water Reactor and the (PWR), Pressurized Water Reactor. The BWR-cooled water is turned into steam within the nuclear reactor itself and run to the turbine outside the containment area. The PWR-cooled water is pressurized in excess of 100 bar and there is no water boiling process inside the reactor. It is then run to the turbine and through a cooling process for recirculation.

Solar Power Steam Generators

Solar power steam generators are the cleanest source of boiling water. The water is run through tubes inside a solar panel. The sun heats the water and then the water runs through a steam turbine, creating electricity. There are no waste products and no pollution.

Working of Steam Generators

Steam generators are used to harness energy liberated as heat in a wide variety of processes and convert it into a form which is more useful, such as mechanical and electrical energy. The heat used is typically produced intentionally for the production of electricity or is captured as a byproduct of some other industrial process. The immediate source of the heat is usually dirty, such as the radioactive fuel in a nuclear power plant, so the first step of steam power generation is to transfer that heat into clean water with a heat exchanger. This is done by having the heat source elevate the temperature of an exchange medium, such as an oil, that is circulated in a closed circuit. The oil in turn heats a water reservoir without contaminating it.

Steam Generation

The hot oil is circulated through a water bath to generate steam. There are several different geometrical schemes for doing this, but the principle remains the same. The heating liquid is diverted into many smaller pipes to increase its surface contact with the water and facilitate rapid heat exchange and steam production. The steam produced in modern nuclear and coal power plants is often at supercritical conditions, or above the critical point in water's phase diagram (374 degrees Celsius and 22 MPa).

Conversion of Heat to Electricity

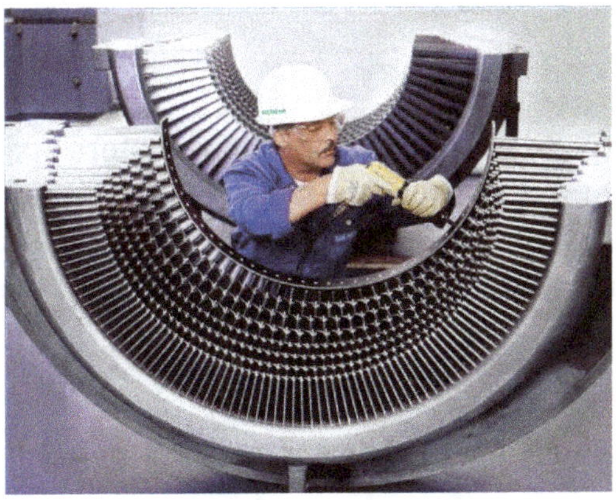

Supercritical steam is overloaded with energy. Steam's energy is converted into mechanical energy by forcing it through a steam turbine. The high pressure of the steam pushes on the many angled blades of the turbine, causing the shaft to rotate. This mechanical energy is converted to electricity by using the power form the rotating shaft to turn an electrical generator. The turbine being built in the image may generate up to 65 megawatts of electricity.

Boiler Efficiency

Boiler efficiency is a combined result of efficiencies of different components of a boiler. A boiler has many sub systems whose efficiency affects the overall boiler efficiency. Couple of efficiencies which finally decide the boiler efficiency are:

1. Combustion efficiency
2. Thermal efficiency

Apart from these efficiencies, there are some other losses which also play a role while deciding the boiler efficiency and hence need to be considered while calculating the boiler efficiency.

Combustion Efficiency

The combustion efficiency of a boiler is the indication of burner's ability to burn fuel. The two parameters which determine the burner efficiency are unburnt fuel quantities in exhaust and excess oxygen levels in the exhaust. As the amount of excess air is increased, the quantity of unburnt fuel in the exhaust decreases. This results in lowering the unburnt fuel losses but elevating the enthalpy losses. Hence, it is quite important to maintain a balance between enthalpy losses and un burnt losses. Combustion efficiency also varies with the fuel being burnt. Combustion efficiency is higher for liquid and gaseous fuels than for solid fuels.

Thermal Efficiency

The thermal efficiency of a boiler specifies the effectiveness of the heat exchanger of the boiler which actually transfers the heat energy from fireside to water side. Thermal efficiency is badly affected by scale formation/soot formation on the boiler tubes.

Direct and Indirect Boiler Efficiency

The overall boiler efficiency depends on many more parameters apart from combustion and thermal efficiencies. These other parameters include ON-OFF losses, radiation losses, convection losses, blow down losses etc. In actual practice, two methods are commonly used to find out boiler efficiency, namely direct method and indirect method of efficiency calculation.

Direct Efficiency

This method calculates boiler efficiency by using the basic efficiency formula:

$$\eta = (\text{Energy output})/(\text{Energy input}) \times 100$$

In order to calculate boiler efficiency by this method, we divide the total energy output of a boiler by total energy input given to the boiler, multiplied by hundred.

Calculation of direct efficiency:

$$E= [Q (H-h)/q*GCV]*100$$

Where,

- Q= Quantity of steam generated (kg/hr)
- H= Enthalpy of steam (Kcal/kg)
- h= Enthalpy of water (kcal/kg)
- GCV= Gross calorific value of the fuel

Indirect Efficiency

The indirect efficiency of a boiler is calculated by finding out the individual losses taking place in a boiler and then subtracting the sum from 100%. This method involves finding out the magnitudes of all the measurable losses taking place in a boiler by separate measurements. All these losses are added and subtracted from 100% to find out the final efficiency. Blow down valve is kept closed during the procedure. This method should be implemented as per the norms provided in BS845 standards. The losses calculated include stack losses, radiation losses, blowdown losses etc.

Comparison of Direct and Indirect Efficiency

Both the methods of finding out boiler efficiency mentioned above have some advantages and some disadvantages associated with them. The greatest advantage of indirect method is that it also speaks about the sources of losses. By finding out indirect efficiency, one can come to know where the losses are increased and can be reduced. On the other hand, direct efficiency values are closer to reality as compared to indirect efficiency on account of uncovered losses such as radiation losses, ON-OFF losses etc. But direct efficiency can only tell us about the magnitude of overall loss. No information about individual losses and their magnitudes is conveyed from direct efficiency calculation. There always exists some difference in the values of direct and indirect efficiencies. Indirect efficiency is measured at a particular time whereas direct efficiency is measured over a period of time and hence, losses on account of fluctuating loads, boiler on-off etc. are also taken into consideration.

Real Time Efficiency Monitoring

Boiler efficiency does not remain fixed and large variations from ideal values take place during the course of operation. Shifting to real time efficiency monitoring can improve the boiler efficiency significantly depending upon the boiler type and actual conditions at site. In nutshell, monitoring and maintaining the boiler efficiency for the overall operational life of the boiler is a must to cut fuel bills and reduce carbon footprint.

Boiler Selection Considerations

Six criteria should be considered when selecting a boiler to meet the application needs. The criteria are:

1. Codes and standards requirements
2. Steam or Hot Water
3. Boiler load
4. Number of boilers
5. Performance considerations
6. Special considerations

Codes and Standards

There are a number of codes and standards, laws, and regulations covering boilers and related equipment that should be considered when designing a system. Regulatory requirements are dictated by a variety of sources and are all focused primarily on safety. For more information on how the various rules affect boiler selection and operation, you may want to contact your local Cleaver-Brooks authorized representative. Here are some key rules to consider:

- The boiler industry is tightly regulated by the American Society of Mechanical Engineers (ASME) and the ASME Codes, which governs boiler design, inspection, and quality assurance. The boiler's pressure vessel must have an ASME stamp. (Deaerators, economizers, and other pressure vessels must also be ASME stamped).
- The insurance company insuring the facility or boiler may dictate additional requirements. Boiler manufacturers provide special boiler trim according to the requirements of the major insurance companies. Special boiler trim items usually pertain to added safety controls. Some industries, such as food processing, brewing, or pharmaceuticals, may also have additional regulations that have an impact on the boiler and the boiler room.
- A UL, ULC, cUL, CSA or CGA listing, or Canadian Registration Number (CRN) may be required. State, local, or provincial authorities may require data on the boiler controls or basic design criteria.
- Most areas have established a maximum temperature at which water can be discharged to the sewer. In this case, a blowdown separator aftercooler is required.
- Most state, local or provincial authorities require a permit to install and/or operate a boiler. Additional restrictions may apply in non-attainment areas where air quality does not meet the national ambient air quality standards and emission regulations are more stringent.
- For all new boilers with inputs over 10 MMBtu/hr, U.S. Federal emission standards apply, including permitting and reporting procedures.

- Limits on fuel sulfur content are frequently set at 0.5% maximum.
- A full-time boiler operator may be required. Operator requirement depends on the boiler's size, pressure, heating surface or volume of water. Boilers can be selected which minimize the requirements, either by falling under the requirements and being exempt or with special equipment that gives the operator more freedom in the facility.
- Most states or provinces require an annual boiler inspection. There may be other requirements on piping as well.

Steam or Hot Water

Now that you have a general overview of the types of boilers and code and standards requirements, it's time to look at the facility's application in order to see how the boiler will be used. Keep in mind, the primary purpose of the boiler is to supply energy to the facility's operations - for heat, manufacturing process, laundry, kitchen, etc. The nature of the facility's operation will dictate whether a steam or hot water boiler should be used.

Hot water is commonly used in heating applications with the boiler supplying water to the system at 180 °F to 220 °F. The operating pressure for hot water heating systems usually is 30 psig to 125 psig. Under these conditions, there is a wide range of hot water boiler products available. If system requirements are for hot water of more than 240 °F, a high temperature water boiler should be considered.

Steam boilers are designed for low pressure or high pressure applications. Low pressure boilers are limited to 15 psig design, and are typically used for heating applications. High pressure boilers are typically used for process loads and can have an operating pressure of 75 to 700 psig. Most steam boiler systems require saturated steam.

Steam and hot water boilers are defined according to design pressure and operating pressure. Design pressure is the maximum pressure used in the design of the boiler for the purpose of calculating the minimum permissible thickness or physical characteristics of the pressure vessel parts of the boiler. Typically, the safety valves are set at or below design pressure. Operating pressure is the pressure of the boiler at which it normally operates. The operating pressure usually is maintained at a suitable level below the setting of the pressure relieving valve(s) to prevent their frequent opening during normal operation.

Some steam applications may require superheated steam. It should be noted that superheated steam has a high enthalpy, so there is more energy per pound of steam and higher (drier) steam quality. One example of an application where superheated steam may be required is with a steam turbine. The turbine's blades require very dry steam because the moisture can destroy the blades. When very high pressure or superheated steam is required, an industrial watertube boiler should be selected.

System Load

In addition to the system load considerations provided in this part, many excellent reference manuals are available to help further define specific load details and characteristics. For more information, refer to the ABMA Firetube Engineering Guide, the ASHRAE Handbook, or contact your local Cleaver-Brooks authorized representative.

System load is measured in either Btus or pounds of steam (at a specific pressure and temperature). When discussing the system load, we will include references to both steam and hot water. However, not all situations or criteria apply to both. It would be nearly impossible to size and select a boiler(s) without knowing the system load requirements. Knowing the system load provides the following information:

- The boiler(s) capacity, taken from the maximum system load requirement.
- The boiler(s) turndown, taken from the minimum system load requirement.
- Conditions for maximum efficiency, taken from the average system load requirement.

Determining the total system load requires an understanding of the type(s) of load in the system. There are three types of loads: heating, process, and combination.

Heating Load

A heating load is typically low pressure steam or hot water, and is relatively simple to define because there is not a great deal of instantaneous changes to the load. And, once a heating load is computed, the number can easily be transferred into the equipment size requirements. A heating load is used to maintain building heat. Cooling loads, using steam to run an absorption chiller, also are included when computing a heating load. Characteristics of a heating load include large seasonal variations but small instantaneous demand changes. The boiler should be sized for the worst possible weather conditions, which means that true capacity is rarely reached.

Process Load

A process load is usually a high pressure steam load. A process load pertains to manufacturing operations, where heat from steam or hot water is used in the process. A process load is further defined as either continuous or batch. In a continuous load, the demand is fairly constant - such as in a heating load. The batch load is characterized by short-term demands. The batch load is a key issue when selecting equipment, because a batch-type process load can have a very large instantaneous demand that can be several times larger than the rating of the boiler. For example, based on its size, a heating coil can consume a large amount of steam simply to fill and pressurize the coil. When designing a boiler room for a process load with instantaneous demand, a more careful boiler selection process should take place.

Combination Load

Many facilities have a mixture of loads - different types of process loads and combinations of heating and process loads. The information just given on heating and process loads should be taken into consideration when dealing with a combination load.

Defining Load Variations

Loads vary and a power plant must be capable of handling the minimum load, the maximum load, and any load variations. Boiler selection is often dictated by the variation in load demand, rather than by the total quantity of steam or hot water required.

There are three basic types of load variations: seasonal, daily, and instantaneous:

1. Seasonal Variations: For a heating system, seasonal variations can mean no demand in the summer, light demand in the fall and spring, and heavy demand in the winter. Manufacturing operations often have seasonal variations, because the demand for production may vary. When selecting boiler equipment, the minimum and maximum load for each season should be determined.

2. Daily Variation: Daily variation can occur due to variations in the work hours, or the heat required at various times of the day or weekend. Minimum and maximum seasonal variations mentioned earlier may already reflect these changes if they occur daily. If not, the minimum and maximum daily loads should be included.

The seasonal and daily variations define the size of the load that the boiler(s) must handle. Seasonal and daily variations also help define the number of boilers and turndown requirements.

3. Instantaneous Demand: Instantaneous demand is a sudden peak load change that is usually of short duration. These types of loads are sometimes hidden. Many machines or processes are rated in pounds of steam per hour or Btu/hr as running loads, under balanced operating conditions, and there is no recognition given to "cold startup," "peak" or "pickup loads." The instantaneous load demand is important to consider when selecting a boiler to ensure that these load variations are taken into account. If the instantaneous demand is not included in the system load calculations, the boiler(s) may be undersized.

Load Tracking

Load tracking is the ability of a boiler to respond to changes in steam or hot water demand. Most often associated with process loads, load tracking focuses on the boiler's ability to supply a constant volume of steam at the required pressure.

The ability of the boiler to track a variable load depends on the boiler type, burner turndown capability, feedwater valve control, and combustion control design. If the analysis of the load shows highly variable load conditions, a more complex control package may be necessary. This type of control is achieved with sophisticated boiler management systems.

If the application has instantaneous load demands, whereby a large volume of steam is required for a short period of time, a boiler with a large energy storage reserve, such as a firetube, should be considered. If the application dictates large variances in load demand, where the load swings frequently for long periods of time, the best choice is probably a watertube type boiler, because it contains less water and can respond to the variances more rapidly.

In all cases, operation of the burner should be taken into account in selecting a boiler(s) to meet system demand. The burner will require proper operating controls that can accurately sense the varying demands and be capable of the turndown requirements. The boiler feedwater valve and control design are also critical if load swings are expected.

Number of Boilers

Back-up Boilers

When selecting the boiler(s), consideration should be given to back-up equipment to accommodate future expansion, emergency repairs, and maintenance. There are a number of considerations for a backup boiler.

Type of Load

Heating systems and non-critical loads that do not result in a sudden loss of production generally have little or no backup. While this is not recommended, it is still common practice. These types of applications rely on the ability to make repairs quickly to reduce downtime. The risk involved in having no backup is a total loss of heat when the boiler is not in service.

When process or heating loads use multiple boilers during peak times, and one boiler during most other times, the availability of an additional boiler to provide full backup during maximum demand should be considered.

In applications with critical steam or hot water requirements, laws or codes may require a backup. Even if laws or codes do not require a backup, there are many cases where the operation cannot tolerate downtime. For example, a hotel uses hot water 24 hours a day, seven days a week. During periods of maintenance or in an emergency, a backup boiler is required.

Downtime

Another way to determine whether a backup boiler is a wise decision is to compute the cost of downtime to the owner or the user, as shown in the following three examples:

- A chemical company manufactures dry cell battery compound in a batch process. The process temperature must be maintained within 2 degrees. The boiler shuts down on a flame failure. They have 20 minutes to recover steam or the batch is scrap. The value of the product is $250,000.
- A Midwestern insurance company building has comfort heat supplied by one boiler. There are over 2000 workers in the building. The boiler shuts down due to a failed gas valve. Outside, it's 11 °F. Inside, the temperature continues to drop and, at 1:30 in the afternoon, all 2,000 workers are sent home.
- A meat processing company makes its entire packaged ham line in a Southern plant. It operates 24 hours a day, every day. A single boiler provides heat for curing, sterilizing, and cleaning. The boiler goes down due to a lack of feedwater. Each hour of steam loss results in four hours of lost production.

Boiler Turndown

Boiler turndown is the ratio between full boiler output and the boiler output when operating at low fire. Typical boiler turndown is 4:1. For example, a 400 horsepower boiler, with a 4:1 turndown burner, will modulate down to 100 horsepower before cycling off. The same boiler with a 10:1 turndown burner will modulate down to 40 horsepower.

The ability of the burner to turn down reduces frequent on and off cycling. Fully modulating burners are typically designed to operate down to 25% of rated capacity. At a load that is 20% of the rated capacity, the boiler will turn off and cycle frequently.

A boiler operating at low load conditions can cycle as frequently as 12 times per hour, or 288 times per day. With each cycle, pre- and post-purge air flow removes heat from the boiler and sends it out the stack. The energy loss can be eliminated by keeping the boiler on at low firing rates. Every time the boiler cycles off, it must go through a specific start-up sequence for safety assurance. It requires about one to two minutes to place the boiler back on line. And, if there's a sudden load demand, the start-up sequence cannot be accelerated. Keeping the boiler on line assures the quickest response to load changes. Frequent cycling also accelerates wear of boiler components. Maintenance increases and, more importantly, the chance of component failure increases. As discussed earlier, boiler(s) capacity requirement determined by many different types of load variations in the system. Boiler over-sizing occurs when future expansion and safety factors are added to assure that the boiler is large enough for the application. If the boiler is oversized, the ability of the boiler to handle minimum loads without cycling is reduced. Therefore, capacity and turndown should be considered together for proper boiler selection to meet overall system load requirements.

Three important considerations pertain to fuels, emissions, and efficiency. All three have important impact on boiler performance, and can affect long-term boiler operating costs.

Fuels

Remember, from an operating perspective, fuel costs typically account for approximately 10% of a facility's total operating budget. Therefore, fuel is an important consideration. Normally, the fuels of choice are natural gas, propane, or light oil. Increasingly stringent emission standards have greatly reduced the use of heavy oil and solid fuels such as coal and wood. Of the fossil fuels, natural gas burns cleanest and leaves less residue; therefore less maintenance is required.

It can be advantageous to supply a boiler with a combination burner that can burn two fuels independently - for example, oil or natural gas. A combination burner allows the customer to take advantage of "peak time" rates, which substantially reduces the costs of a therm of gas when operating "off peak" by merely switching to the back up fuel. Dual fuel capability also is beneficial if the primary fuel supply must be shut down for safety or maintenance reasons.

Some waste streams can be used as fuel in the boiler. In addition to reducing fuel costs, firing an alternate fuel in a boiler can greatly reduce disposal costs. Waste streams are typically used in combination with standard fuels to ensure safe operation and to provide additional operating flexibility.

Emissions

Emission standards for boilers have become very stringent in many areas, because of the new clean air regulations. The ability of the boiler to meet emission regulations depends on the type of boiler and burner options.

Efficiency is used in the measure of economic performance of any piece of equipment. In the boiler industry, there are four common definitions of efficiency, but only one true measurement. Following are the definitions and how to measure efficiency.

Combustion Efficiency

Combustion efficiency is the effectiveness of the burner only and relates to its ability to completely burn the fuel. The boiler has little bearing on combustion efficiency. A well- designed burner will operate with as little as 15 to 20% excess air, while converting all combustibles in the fuel to thermal energy.

Thermal Efficiency

Thermal efficiency is the effectiveness of the heat transfer in a boiler. It does not take into account boiler radiation and convection losses - for example, from the boiler shell, water column piping, etc.

Boiler Efficiency

The term "boiler efficiency" is often substituted for combustion or thermal efficiency. True boiler efficiency is the measure of fuel-to-steam efficiency.

Fuel-to-Steam Efficiency

Cleaver-Brooks guaranteed boiler efficiencies are fuel-to-steam efficiencies.

Fuel-to-steam efficiency is the correct definition to use when determining boiler efficiency. Fuel-to-steam efficiency is calculated using either of two methods, as prescribed by the ASME Power Test Code, PTC 4.1. The first method is inputoutput. This is the ratio of Btu output divided by Btu input x 100.

The second method is heat balance. This method considers stack temperature and losses, excess air levels, and radiation and convection losses. Therefore, the heat balance calculation for fuel-to-steam efficiency is 100 minus the total percent stack loss and minus the percent radiation and convection losses.

Stack Temperature and Losses

Stack temperature is the temperature of the combustion gases (dry and water vapor) leaving the boiler. A well-designed boiler removes as much heat as possible from the combustion gases. Thus, lower stack temperature represents more effective heat transfer and lower heat loss up the stack. The stack temperature reflects the energy that did not transfer from the fuel to steam or hot water. Stack temperature is a visible indicator of boiler efficiency. Any time efficiency is guaranteed, predicted stack temperatures should be verified.

Stack loss is a measure of the amount of heat carried away by dry flue gases (unused heat) and the moisture loss (product of combustion), based on the fuel analysis of the specific fuel being used, moisture in the combustion air, etc.

Excess Air

Excess air provides safe operation above stoichiometric conditions. A burner is typically set up with 15 to 20% excess air. Higher excess air levels result in fuel being used to heat the air instead of transferring it to usable energy, increasing stack losses.

Radiation and Convection Losses

Radiation and convection losses will vary with boiler type, size, and operating pressure. The losses are typically considered constant in Btu/hr, but become a larger percentage loss as the firing rate decreases.

Boiler design factors that also impact efficiencies of the boiler are heating surface, flue gas passes, and design of the boiler and burner package.

Heating Surface

Heating surface is one criterion used when comparing boilers. Boilers with higher heating surface per boiler horsepower tend to be more efficient and operate with less thermal stress. Many packaged boilers are offered with five square feet of heating surface per boiler horsepower as an optimum design for peak efficiency.

Flue Gas Passes

The number of passes that the flue gas travels before exiting the boiler is also a good criterion when comparing boilers. As the flue gas travels through the boiler it cools and, therefore, changes volume. Multiple pass boilers increase efficiency because the passes are designed to maximize flue gas velocities as the flue gas cools.

Integral Boiler/Burner Package

Ultimately, the performance of the boiler is based on the ability of the burner, the boiler, and the controls to work together. When specifying performance, efficiency, emissions, turndown, capacity, and excess air all must be evaluated together. The efficiency of the boiler is based, in part, on the burner being capable of operating at optimum excess air levels. Burners not properly designed will produce CO or soot at these excess air levels, foul the boiler, and substantially reduce efficiency. In addition to the boiler and burner, the controls included on the boiler (flame safeguard, oxygen trim, etc.) can enhance efficiency and reduce overall operating costs for the customer. A true packaged boiler design includes the burner, boiler, and controls as a single, engineered unit.

Replacement Boilers

If the boiler is to be placed in an existing facility, there are a number of considerations:

- Floor space required.
- Total space requirements.
- Access space for maintenance.

- Size and characteristics of the boiler to be replaced, including location of existing piping, the boiler stack and utilities.
- Boiler weight limitations.
- With little or no access to the boiler room, some bent-tube type boilers can be carried into the boiler room in sections or pieces and easily assembled, with no welding required.
- Electric boilers should also be considered, especially since they do not require a stack.
- Vertical firetube boilers have a small floor space requirement.

Boiler Mountings and Accessories

The various boiler mountings and accessories that are used in steam boilers are water level indicator, pressure gauge, safety valves, stop valve; blow off cock, feed check valve, fusible plug, air pre-heater, super heater, economiser and feed pump. The boiler mounting and accessories are used in steam boilers for its proper, efficient and satisfactory working. In this topic we will discuss about the functions of each of them.

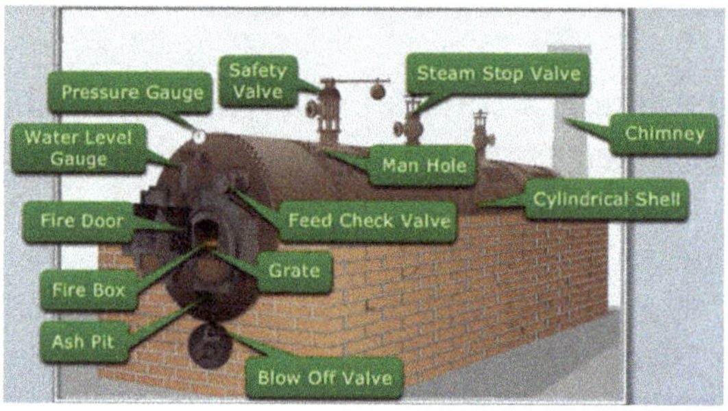

The different boiler mountings and accessories that are installed on the steam boiler are as follows:

Boiler Mountings

1. Water level indicator (Water level gauge).
2. Pressure gauge.
3. Safety valves.
4. Stop valve.
5. Blow off cock (Blow off valve).
6. Feed check valve.

Boiler Accessories

1. Air pre-heater.

2. Super heater.
3. Economiser.
4. Feed pump.

The function of various boiler mountings and accessories are:

Water Level Indicator

- It is fitted in front of the boiler and generally present two in number.
- It is used to indicate the water level inside the boiler. It shows the instantaneous level of water that is present inside the steam boiler which is necessary for its proper working.

Pressure Gauge

- It is also present in front of the boiler.
- It is used to measure the pressure of the steam inside the boiler.
- The pressure gauges generally used are of Bourden type.

Safety Valves

- Safety valves are attached to the steam boiler chest.
- It is used to prevent explosion due to excessive internal pressure. When the internal pressure inside the boiler exceeds its working pressures than the safety valves blow off the steam and maintains the internal pressure.
- Generally two safety valves are present on a boiler.

Stop Valve (Steam Stop Valve)

- It is usually fitted on the highest part of the boiler with the help of a flange.
- The main function of the stop valve is:
    - To control the flow of steam from the boiler to the main steam pipe.
    - To completely shut off the steam supply when required.

Blow Off Valve

- It is fitted at the bottom of the boiler drum.
- The functions of blow off cock is:
    - To empty the boiler whenever required.
    - To discharge the scale, mud and sediments which gets collected at the bottom of the boiler.

Feed Check Valve

- It is non-return valve and fitted to a screwed spindle to regulate the lift.
- It is fitted to the shell slightly below the normal water level of the boiler. A boiler must have its spindle lifted before the pump is started.
- It regulates the supply of water which is pumped into the boiler by feed pump.

Fusible Plug

- It is fitted to the crown plate of the furnace or firebox.
- Its function is to extinguish fire in the furnace when the water level in the boiler falls to an unsafe limit. This avoids the explosion that may takes place because of the overheating of the furnace plate.

Boiler accessories: The boiler accessories are the integral parts of the boiler. They are used in the boiler to improve its efficiency.

Air Preheater

- It is used to recover heat from the exhaust gases.
- It is installed between the economiser and the chimney.

Super Heater

- It is placed in the path of hot flue gases from the furnace.
- A super heater is an important accessory used in the boiler. Its main function is to increase the temperature of saturated steam without raising its pressure.

Economiser

- It is used to heat the feed water by the utilization of heat from the hot fuel gases before it leaves the chimney.
- A economiser improves the economy of the steam boilers.

Feed Pump

- It is used to deliver water to the boiler.

Locomotive Boiler

A locomotive boiler is a device which is used to create steam from water by using heat energy. It is a horizontal drum axis, multi-tubular, natural circulation, artificial draft, forced circulation, mobile, medium pressure, solid fuel fired fire tube boiler that has an internal fire furnace. It is capable of producing high steam rate and hence it is used for railway locomotive engines and in marines.

Construction of Locomotive Boiler

A locomotive boiler comprises of various parts, and going through each part will make us understand the construction of a locomotive boiler. These parts are mentioned below:

- Fire hole: This hole is used to insert solid fuel and ignite it in the furnace. It is located at the rear end of a boiler.
- Fire box: In this box, burning of fuel takes place.
- Grate: It is the platform on which fuel is kept and ignited.
- Firebrick arch: It is placed over the grate in an inclined position to stop the entry of dust, ash, and burnt fuel particles into fire tubes. It allows the movement of fuel gases in such a way that they move in a definite path before going into the fire tubes of the boiler.
- Boiler tubes: Through these tubes, hot flue gases pass and exchange the heat with the surrounding.
- Smokebox: The burnt fuel smoke, after passing through fire tubes, get collected in the smoke box and then released into the environment through the chimney.
- Blast pipe: It is located above the steam engine and exhaust steam is passed through this pipe. It creates an artificial draft that pushes the smoke out through the chimney. Also, it creates suction for hotter flue gases in order to move forward through the fire tubed.

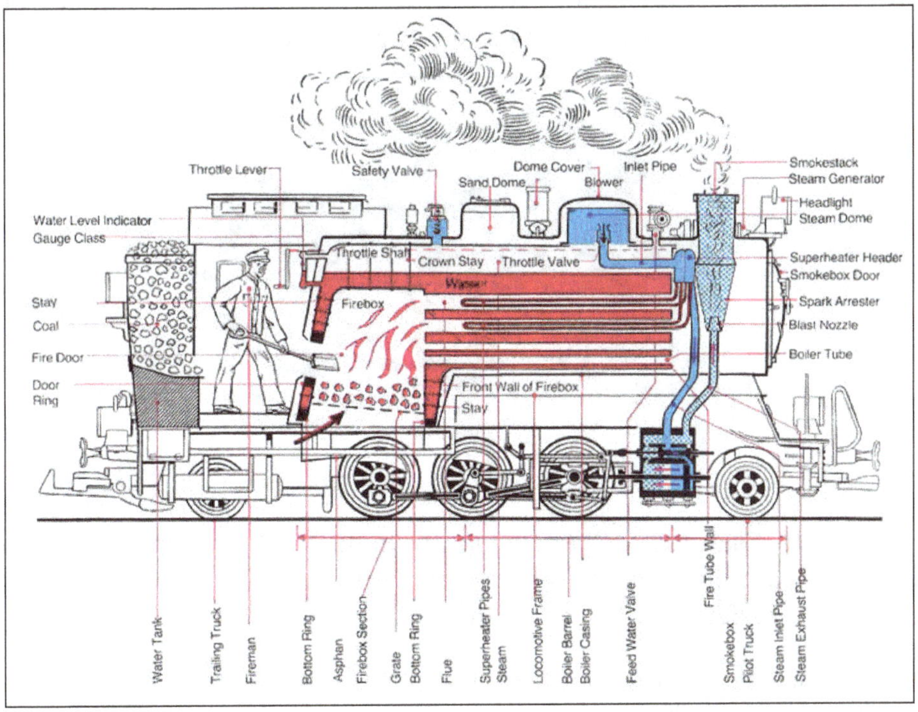

- Steampipe: Steam is passed through the steam pipe. There are two types of steam pipes;

one is used between superheater and dome which is called as main steam pipe and another is used between superheater end exit and steam engine.

- Superheater: The steam is superheated with the help of superheater at the desired temperature before entering into the steam engine cylinder.
- Superheater element pipes: These pipes are present in the superheater through which steam moves and gets superheated.
- Dome: It is used to regulate the steam generated through the steam pipe. It is located on the top.
- Regulator valve: During the superheating process, this valve regulates the steam through the main steam pipe.
- Safety valve: For a safe working steam pressure this valve is used. When pressure is more than the safety level, it blows off and prevents the boiler from blasting.
- Superheater header: To accept the steam from the steam pipe, superheater header is used.
- Chimney: To send out the exhaust smoke and gases from the boiler, a chimney is used which has a very small length.

Locomotive Boiler: Construction, Working and Application

The locomotive boiler works with the insertion of solid fuel on the grate and then igniting it through the fire hole. As the fuel is burnt, it produces hot flue gases. These hot flue gases are moved through a definite path by firebrick arch into the long fire tubes and heat the water in the surrounding. As the heating continues, the water starts converting into saturated steam which gets collected at the top.

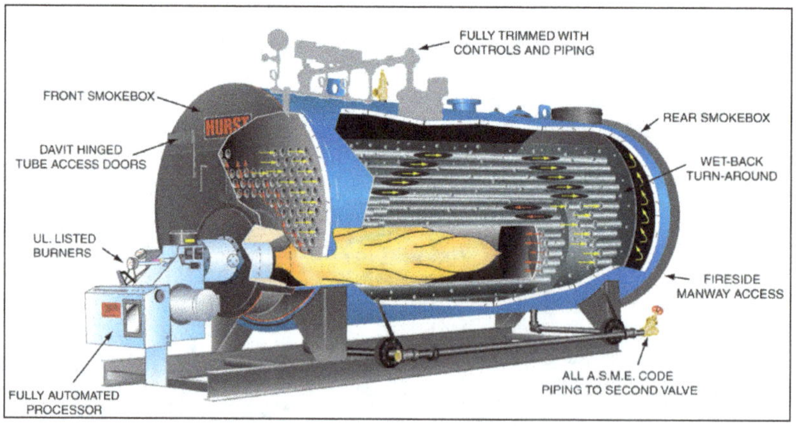

Then, from the dome, this saturated steam gets into the main steam pipe via regulator valve and reaches into the superheater header. The steam enters into the pipes of superheater element to get superheated and then transferred to the steam pipe of smoke box.

The steam from the superheater then enters into the cylinder which contains a piston. With the pressure exerted by the steam, the piston is moved. The piston is connected to the wheels of a locomotive, thus power is directly transferred to the wheels and they start rotating.

The burnt gases and smoke are transferred to the smoke box through fire tubes and exhaust steam from the cylinder is transferred to the blast pipe. The exhaust steam pushes out the smoke from the boiler through the chimney, by coming out from the blast pipe. However, the smoke doesn't move out from the boiler on its own, therefore artificial draft is made by exhaust steam that comes out of the steam engine. This artificial draft pushes the smoke out and simultaneously creates suction for hot flue gases.

Advantages

- It can meet the fluctuating need of steam and can be ported easily.
- It provides high steam production rate.
- Its operation is easy and it is cost effective along with a compact size.

Disadvantages

- Over time, it gets corroded and scale is formed.
- Gets overheated when works under heavy load.
- Cleaning of some water spaces is not easy.

Application of Locomotive Boiler

The locomotive boiler has low efficiency and cannot be used under heavy load because of its over-heating problem. They are commonly used in the railways and marines. In addition, they can be used in a portable steam engine, steam rollers, traction engines, and in other steam road vehicles.

References

- What-is-steam-boiler-working-principle-types-of-steam-boilers: elprocus.com, Retrieved 25 May, 2019
- List-6863907-types-steam-generators: sciencing.com, Retrieved 26 June, 2019
- Steam-generators-work-5095506: sciencing.com, Retrieved 27 July, 2019
- Boiler-Efficiency-Introduction-and-Method-of-Calculation, Boilers, steampedia, Knowledge: for besmarshall.com, Retrieved 28 August, 2019
- Locomotive-boiler-working-application: engineeringinsider.org, Retrieved 29 January, 2019

4
Understanding Draught

Draught is defined as the pressure difference in the boiler system due to which the gas flows from one point to another. Artificial draught, induced draught, balanced draught and steam jet draught are some of the different types of draught. This chapter discusses in detail these types of draught.

Draught is the pressure difference which causes a flow of air or gases from one point to another point in the boiler system. Draught is required in a boiler system mainly due to two reasons:

1. To supply sufficient air for completing the combustion.
2. To remove flue gases from the system after combustion and the heat exchange.

There are two types of draught applied to the boiler system:

1. The natural draught.
2. The forced draught.

The natural draught is always preferred since it does not need any running cost although it has a big initial cost. Natural draught allows natural circulation of air through the boiler system. The natural draughtmainly depends upon the height of the chimney.

We try to calculate the required height of the chimney for a required natural draught for a boiler system. For that, we have to go through two basic equations of gaseous pressure. The equations are:

$$P = \rho gh$$

Where, "P" is the pressure of air or gas, "ρ" is the density of the air or gas, "g" is the gravitational constant, and "h" is the height of the head:

$$PV = mRT$$

Here "V" is the volume of the air or gas, "m" is the mass of the gas or air, "T" is the temperature measured in kelvin scale and "R" is the gas constant.

Equation $PV = mRT$ can be rewritten as:

$$PV = mRT \Rightarrow P = \frac{m}{V}RT \Rightarrow P = \rho RT \Rightarrow \rho = \frac{P}{RT}$$

During combustion process in the furnace, mainly carbon reacts with oxygen (O_2) of the air and forms carbon dioxide (CO_2). The volume of the solid carbon compared to the required air for the reaction is negligible. Because of that, we can consider that the volume of air required for combustion is exactly equal to the volume of the flue gases created after combustion if we assume temperature before and after combustion are same. But this is not the actual case. The air entries in the combustion chamber will gain extra volume after combustion due to combustion temperature. The gained volume of the air will be equivalent to the volume of the flue gases created after combustion:

$$\rho_o = \frac{P}{RT_o}$$

Let us assume, ρ_0 is the density of the air at 0 °C or 273 K, and say it is T_0.

Here, P is the pressure of air at 0 °C or 273 K, that is at T_0 K.

If we keep, the pressure P as constant, the relation between density and temperature of the air or gases can be written as,

$$\rho_o = \frac{P}{RT_o},\ \rho_a = \frac{P}{RT_a},\ \rho_g = \frac{P}{RT_g}$$
$$\Rightarrow \rho_o T_o = \rho_o T_a = \rho_g T_g$$

where, ρ_a and ρ_g is the density of the air at temperature T_a and T_g K respectively.

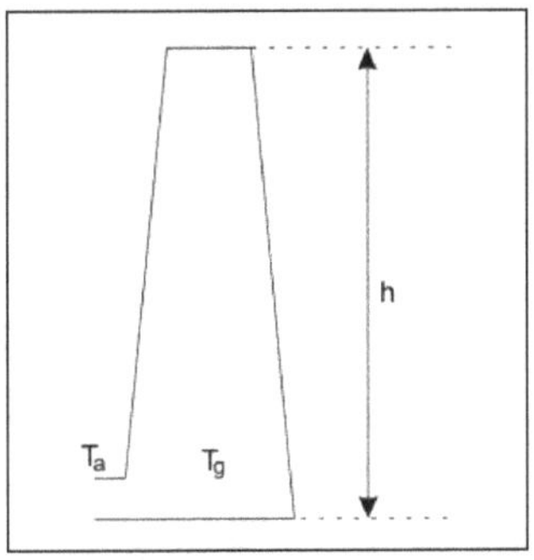

From, equation $P = \rho gh$ and $\rho_o T_o = \rho_o T_a = \rho_g T_g$ we can write the expression of pressure at point

"a" outside the chimney, as:

$$P_a = \rho_a gh = \frac{\rho_a T_o}{T_a} gh$$

The volume of the air at the temperature T_g would be:

$$\rho_g = \frac{P}{RT_g} \text{ and volume } V_g = \frac{m}{\rho_g}$$

Let us assume, m kg of air is required to burn 1 kg of carbon then the density of the flue gas would be:

$$\rho_g' = \frac{m+1}{m/\rho_g} = \frac{m+1}{m}\rho_g = \frac{m+1}{m} \times \frac{\rho_o T_o}{T_g}$$

The pressure of the flue gas inside the cheimny from equation $P = \rho g h$ and $\rho_g' = \frac{m+1}{m/\rho_g} = \frac{m+1}{m}\rho_g = \frac{m+1}{m} \times \frac{\rho_o T_o}{T_g}$, would be: $P_g = \frac{m+1}{m} \times \frac{\rho_o T_o}{T_g} gh$

The pressure difference between outside and inside of the chimney from above equations would be:

$$\Delta p = P_a - P_g = \frac{\rho_o T_o}{T_a} gh - \frac{m+1}{m} \times \frac{\rho_o T_o}{T_g} gh = \left[\frac{1}{T_a} - \frac{m+1}{mT_g}\right]$$

Here, "h" is the minimum height of the chimney to be constructed for the draught ΔP. The flue gas will flow upwards through the chimney due to this pressure difference. So, by calculating this pressure difference one can easily calculate the approximate height of the chimney to be constructed. The pressure difference can be represented as a formula for calculating the height of the chimney for a natural draught.

Artificial Draught

The expression of the pressure difference between inside and outside of a chimney shows that it has a term which is the height of the chimney. To achieve sufficient pressure difference between two points or to attain enough natural draught we have to increase the height of the chimney which is not always practical in the view of cost-effectiveness. In those cases, we have to think about artificial draught instead of natural draught. We can achieve artificial draught, mainly with two methods. One gets produced by steam jet, and forced air produces another. Introducing artificial draught one can significantly reduce the height of the chimney to fulfil the same purpose removing flue gasses to the atmosphere.

Induced Draught

Here, an air fan is attached to the base of the chimney or nearer to it. When the fan rotates, it sucks the flue gases from the boiler furnace system. The suction of flue gases from the furnace producers a pressure difference between the outside air and inside flue gases, and which creates a draught. Due to this draught, fresh air enters the furnace. As the draught gets induced due to the suction of gases, we call the method as induced draught. ID fan or induced draught fan sucks the flue gases from the boiler system and pushes to atmosphere through the chimney along its height. In natural draught, the temperature of the flue gases plays a vital role in traveling the gases through the chimney to the atmosphere. But in the case of induced draught, the temperature of flue gas is not an essential parameter. We can utilize the heat energy of the flue gases as much as possible. In induced draught after extracting almost entire heat from the flue gases, the cold flue gases, we remove the gases to the atmosphere forcefully. So we can achieve the purpose of a very tall chimney with much shorter one which is a significant saving of money for the system.

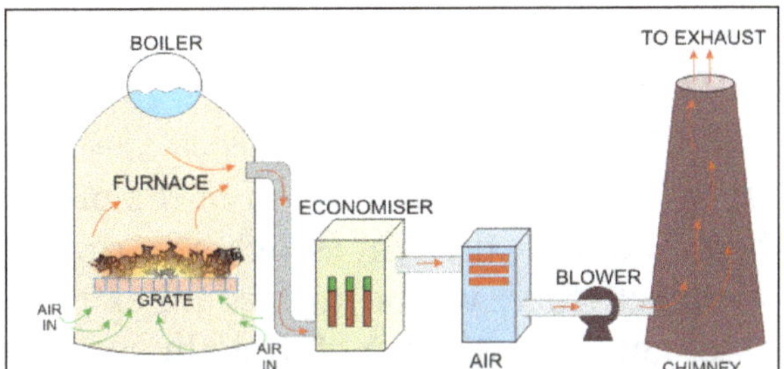

Induced Draught

Forced Draught

Theoretically induced draught and forced draught are more or less same. The only difference is that in induced draught the suction fan is used and in case of forced draughtthe blower fan is used. In case of forced draught system, we fit the blower fan before the coal bed. The blower blows the air from the atmosphere to the coal bed and grate where flue gases get created after combustion. The fresh air (preheated) coming to the furnace pushes the flue gases inward. The flue gases then pass through the economiser, air preheater, etc. toward the stack of the chimney. The forced draught creates positive pressure inside the system. Because of that, we should take special care to protect the system from any leakage otherwise the performance of the system gets affected.

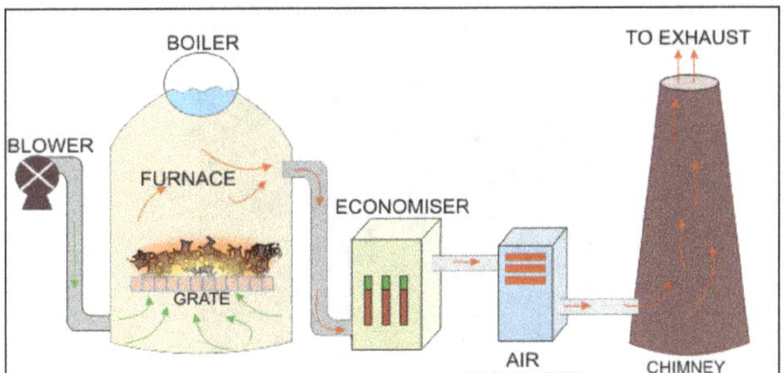

Forced Draught

Balanced Draught

Balanced draught is a combination of forced draught and induced draught. Here we attached blower fans at the entry point of the furnace and induced fans at exit point as well. Here, we utilize benefits of both forced and induced draught. In the balanced draught system, we use forced draught to push the air to the coal, to the grate and later to air preheater. We use the induced draught to withdraw the gases from the economiser and air preheater etc., and then ultimately to discharge the gases to the top of the stack of the chimney.

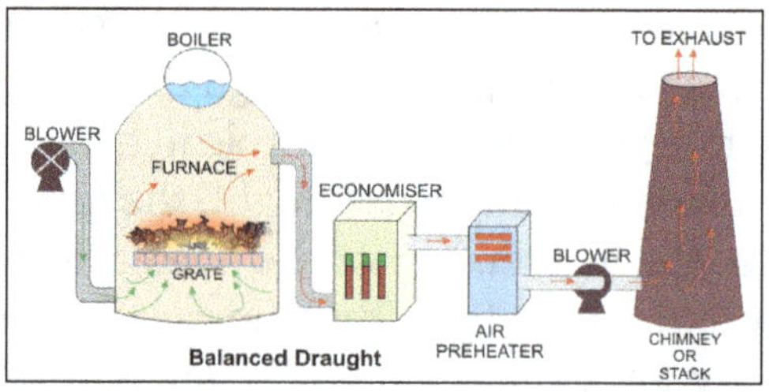

Balanced Draught

Steam Jet Draught

The boiler produces steam. If we utilize, a small portion of the pressurized steam for producing draught in a boiler system then we call the draught as steam jet draught. As the steam, causes the draught, generated in the boiler itself, there will be no need for extra electric power to run draught fans. Steam jet draught is a straight forward form of draught system in a boiler. No demand for additional electricity for running the draught fans hence reducing the cost of the system.

Construction of the system is simple and easy to maintain. Thus, maintenance cost is also low. In steam jet draught system a small portion of generated steam exhausts through a nozzle and the kinetic energy of the jet of high-velocity steam drags the air or flue gases in the boiler system. We can categorize the stream jet draught in two types. One is natural stream jet draught, and another is force steam jet draught.

Forced Stream Jet Draught

In forced stream jet draught, we feed a portion of the stream produced in the boiler to the entry point of the furnace through diffusion pipe. Due to the kinetic energy of steam, there will be a draught at the entry point because of which fresh air gets dragged to the grate and then to the economiser then to preheater then ultimately to the chimney.

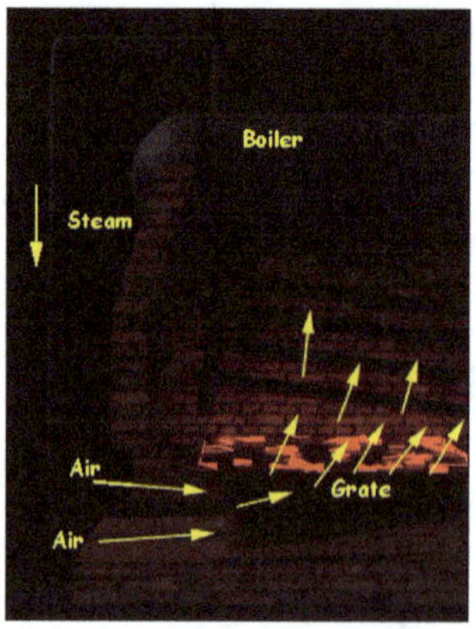

Natural Steam Jet Draught

In natural steam Jet draught, we fit the steam nozzle in a smoke box attached to the bottom portion of the stack. The steam forcefully enters in the smoke box due to which flue gases created in the furnace get dragged in the smoke box because of the resultant draught created the kinetic energy of the steam jet. This process of creating draught in this way is called natural jet steam draught.

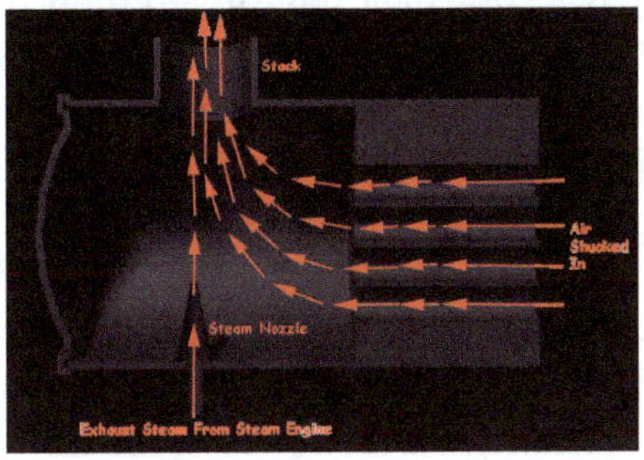

Steam jet draught is simple, economical and occupies very little space or occupies no space. But draught is only possible after steam gets produced which is the main disadvantage of steam Jet draught.

Production of Draught and its Losses

Boiler draught may be defined as the small difference between the pressure of outside air and that of gases within a furnace or chimney at the grate level, which causes flow of air/hot flue gases to take place through boiler.

The draught is necessary to force air through the fuel bed/grate to aid in proper combustion of fuel and to remove the products of combustion i.e. flue gases to the atmosphere after they have given their heat to water being evaporated in boiler. Draught also provides velocity to flue gases and so increases the heat transfer co-efficient in the boiler. Thus draught is essentially required in a boiler and can be produced by a number of methods.

Methods of Production of Draught

The draught may be classified on the basis of how it is produced and is illustrated.

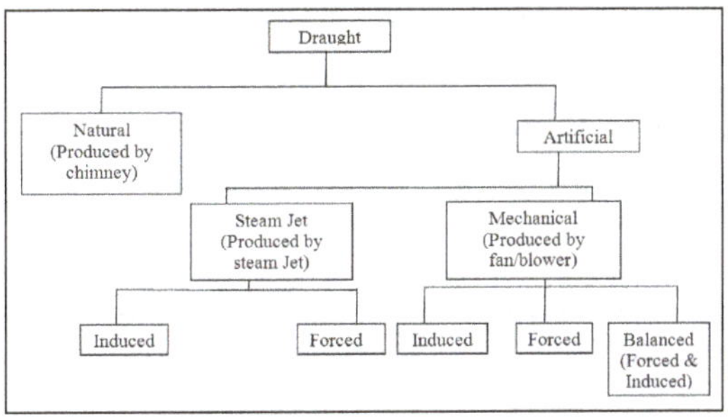

Classification of the draught.

Natural Draught-Chimney

Natural draught is obtained naturally by the use of a chimney. Chimney is a conical shape vertical tubular steel or masonry or concrete structure having a large height as shown in figure. The flue gases after transferring their heat in the boiler are guided by chimney to a considerable height in the atmosphere. As chimney has a large height and is the only outlet from boiler, it remains filled by hot flue gases. These exhaust gases, however have given their heat to water in the boiler, are still hotter than the boiler room air. Due to this, the exhaust gases are lighter in weight than outside air and so lifts up naturally in the chimney and finally escape out in the atmosphere from top of the chimney. As the flue gases lift up in the chimney, more gases from fuel bed flow towards chimney to take their place. During this proces, they first flow through the boiler and do the job of heating of water and produce steam. As flue gases flow through boiler and then to chimney, fresh air from outside naturally enters the boiler and help in burning of fuel and production of hot flue gases. So, in this way, a small pressure difference is naturally created between the base of chimney and the air inlet point of boiler because of density difference between hot flue gases inside the chimney and fresh colder air outside the chimney. This pressure difference is called natural draught, because it is produced naturally or automatically by chimney without any external aid. However the amount of draught produced by chimney is very small and is used in small boilers only. Chimney is also required otherwise to safely remove the harmful poisonous flue gases in the atmosphere at a safe height, where they may not harm the freshness of air on earth.

In this way, a chimney performs one or more of the following functions:

1. It produces draught due to the density difference between the column of hot gases inside the chimney and the cold air outside the chimney.

2. It carries the products of combustion to such a height in the atmosphere, before discharging that they will not be objectionable or hazardous to surroundings.

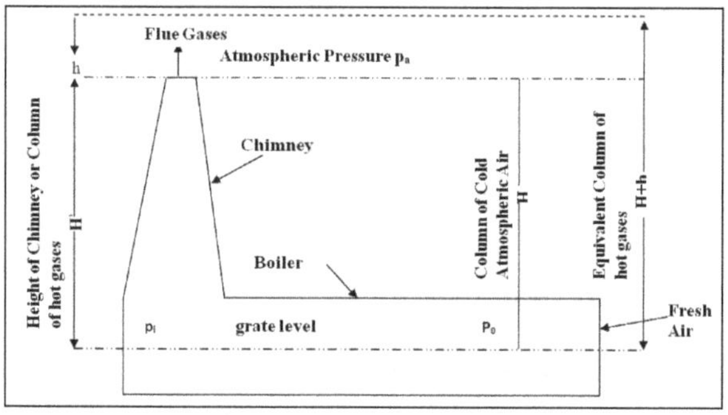

Natural draught system.

Advantages of Mechanical Draught

1. It is more economical and its control is easy.

2. Desired value of draught can be produced by mechanical means which cannot be produced by means of natural draught.

3. It increases the rate of combustion by which low grade fuel can also be used.

4. It reduces the smoke level and increases the heat transfer co-efficient on flue gases side thus increases the thermal efficiency of boiler.

5. It saves the energy and the heat of flue gases can be best utilized by it.

6. In this way, it reduces the fuel consumption and makes boiler operation cheaper.
7. It reduces the height of chimney which now is only controlled by the requirement of pollution norms.

Disadvantages of Mechanical Draught

1. Initial costs of mechanical draught system are high.
2. Running cost is also high due to requirement of electricity but that is easily compensated by the savings in fuel consumption.
3. Maintenance cost is also on higher side.
4. Noise level of boiler is also high due to noisy fan/blower etc.

Table: Comparison between forced draught and induced draught.

Forced Draught	Induced Draught
Fan or blower is placed before grate or situated within the oil burner assembly in case of oil fired boiler.	The fan or blower is placed after the grate or after the flue passages and before chimney.
Pressure inside the flue passages is slightly more than atmospheric pressure.	Pressure inside the flue passages is slightly less than atmospheric pressure.
Due to above reason, there is serious danger of fire to come out and injure someone if any leakage in boiler flue passage take place.	Due to above reason, there is no danger of fire to come out from boiler flue passage in case of a leakage. Hence Induced draught is more safe.
It forces fresh air into the combustion chamber, which helps in burning of fuel and production of hot combustion gases. Due to this force, the hot flue gases are further pushed through the flue passages in boiler.	It sucks the hot flue gases from the combustion chamber through flue passages and then passes on these to economizer, air pre heater and chimney. Due to this suction, fresh air is also sucked in to combustion chamber.
Forced draught fan require less power because it has to handle cold and dense air, the volume per unit mass of which is less.	Induced draught fan requires more power because it has to handle hot flue gases, the volume per unit mass of which is more.
Flow of flue gases through boiler is more uniform.	Flow of flue gases is less uniform.

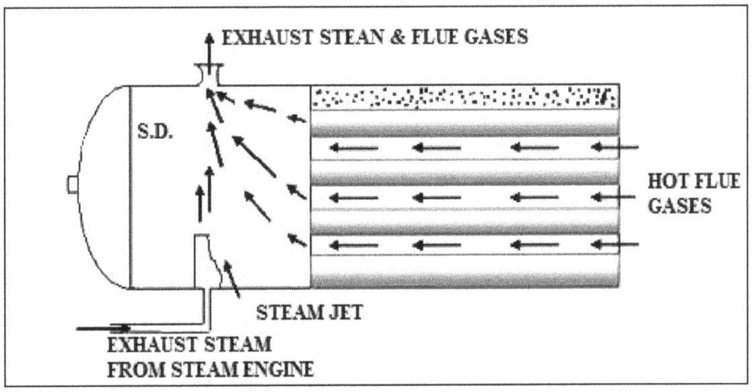

Induced steam jet draught.

Advantages of Artificial Draught over Natural Draught

1. Easy control on combustion of fuel and evaporation of water.
2. Significant increase in capacity or evaporation power of boiler.
3. Increase in fuel burning capacity of grate.
4. Fuel consumption decreases sufficiently.
5. Improvement in efficiency of plant.
6. Reduced chimney height.
7. Prevention of smoke.
8. Low grade fuel can be used.

Draught Losses

The draught is required to overcome the various draught losses in boiler as discussed below:

1. Loss due to the frictional resistance offered by flue gas passage to the flow of flue gases.
2. Loss due to bends in gas flow circuit, which also offer flow resistance.
3. Loss due to friction head in grate, economizer, super heater etc.
4. Loss due to flow resistance offered by chimney.
5. Loss due to imparting some velocity to flue gases, which is required to increase heat transfer in boiler and also to throw away the flue gases from chimney.

5

Steam Nozzle and Injector

A duct through which the velocity of a fluid increases due to drop in pressure is called a steam nozzle. A steam injector is a passive jet pump which produces higher pressure water than the inlet fluid pressure without using external power source. This chapter closely examines the key concepts of steam nozzle and injector to provide an extensive understanding of the subject.

Steam Nozzle

Nozzle is a duct by flowing through which the velocity of a fluid increases at the expense of pressure drop. if the fluid is steam, then the nozzle is called as Steam nozzle.

The flow of steam through nozzles may be takenas adiabatic expansion. The steam possessesa very high velocity at the end of the expansion, and the enthalpy decreases as expansion occurs. Friction exists between the steam and the sides of the nozzle; heat is produced as the result of the resistance to the flow. The phenomenon of super saturation occurs in the steam flow through nozzles. This is because of the time lag in the condensation of the steam during the expansion.

The area of such duct having minimum cross-section is known as throat.

A fluid is called compressible if its density changes with the change in pressure brought about by the flow.

If the density changes very little or does not changes, the fluid is said to be incompressible. Generally the gases and vapors are compressible, whereas liquids are incompressible.

Types of Nozzles

There are three types of nozzles:

1. Convergent nozzle
2. Divergent nozzle
3. Convergent-divergent nozzle

Convergent Nozzle

A typical convergent nozzle is shown in the figure. In a convergent nozzle, the cross sectional area decreases continuously from its entrance to exit. It is used in a case where the back pressure is equal to or greater than the critical pressure ratio.

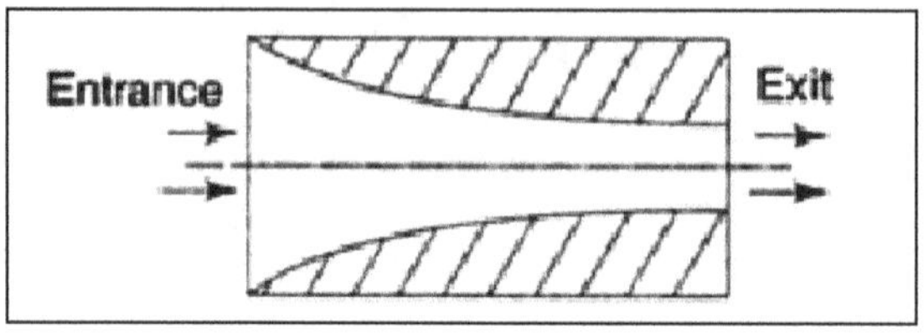

Convergent nozzle.

Divergent Nozzle

The cross sectional area of divergent nozzle increases continuously from its entrance to exit. It is used in a case where the back pressure is less than the critical pressure ratio.

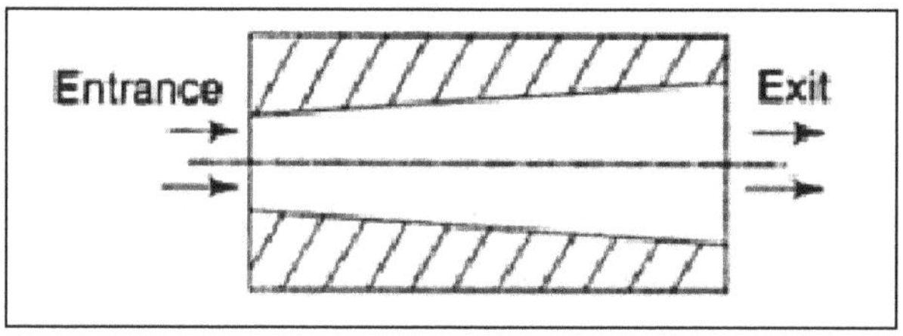

Divergent nozzle.

Convergent – Divergent Nozzle

In this condition, the cross sectional area first decreases from its entrance to the throat and then again increases from throat to the exit. This case is used in the case where the back pressure is less than the critical pressure. Also, in present day application, it is widely used in many types of steam turbines.

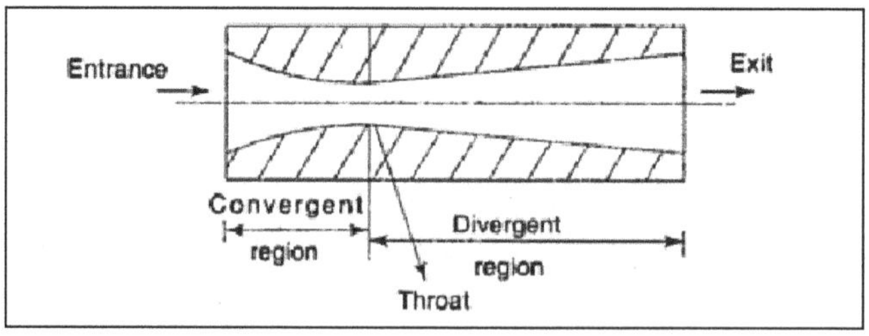

Convergent-Divergent nozzle.

Flow of Steam through Nozzle

Supersaturated flow or metastable flow of in Nozzles. As steam expands in the nozzle, the pressure and temperature in it drop, and it is likely that the steam start condensing when it strikes the saturation line. But this is not always the situation. Due to the high velocities, the time up to which the steam resides in the nozzle is small, and there may not be sufficient time for the needed heat transfer and the formation of liquid droplets due to condensation. As a result, the condensation of steam is delayed for a while. This phenomenon is known as super saturation, and the steam that remains in the wet region without holding any liquid is known as supersaturated steam. The locus of points where condensation occursregardless of the initial temperature and pressure at the entrance of the nozzle is called the Wilson line. The Wilson line generally lies between 4 and 5 percent moisture curves in the saturation region on the h-s diagram in case of steam, and is often taken as 4 percent moisture line. The phenomenon of super saturation is shown on the h-s chart below:

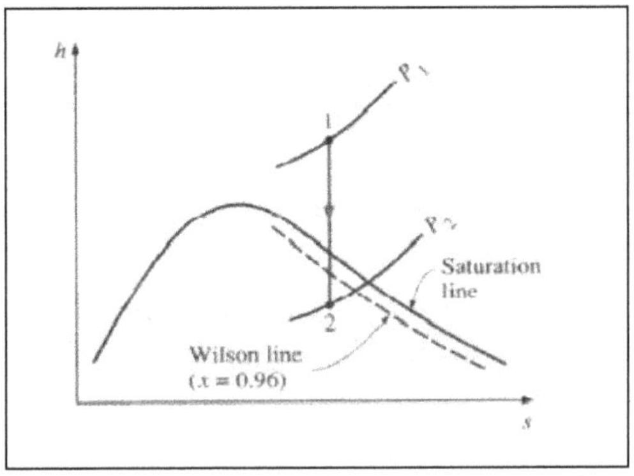

The h-s diagram for the expansion of steam in the nozzle.

Effects of Supersaturation

The following are the effects of supersaturation in a nozzle.

1. The temperature at which the steam becomes supersaturated will be less than the saturation temperature corresponding to that pressure. Therefore, supersaturated steam will havethe density more than that of equilibrium condition which results in the increase in the mass of steam discharged.
2. Supersaturation causesthe specific volume and entropy of the steam to increase.
3. Supersaturation reduces the heat drop. Thus the exit velocity of the steam is reduced.
4. Supersaturation increases the dryness fraction of the steam.

Effect of Friction on Nozzles

1. Entropy is increased.
2. The energy available decreases.
3. Velocity of flow at the throat getdecreased.
4. Volume of flowing steam is decreased.
5. Throat area required to discharge a given mass of steam is increased.

Continuity and Steady Flow Energy Equations

Where m denotes the mass flow rate, v is the specific volume of the steam, A is the area of cross-section and C is the velocity of the steam.

For steady flow of the steam through a certain apparatus, principle of conservation of energy states:

$$h_1 + C_1^2/2 + gz_1 + q = h_2 + C_2^2/2 + gz_2 + w$$

For nozzles, changes in potential energies are negligible, w = 0 and q = 0.

$$h_1 + C_1^2/2 = h_2 + C_2^2/2$$

which is the expression for the steady state flow energy equation.

Suppose a nozzle is used to obtain a supersonic stream starting from low speeds at the inlet, then the Mach number should increase from Ma=0 near the inlet to Ma>1 at the exit. It is clear that the nozzle must converge in the subsonic portion and diverge in the supersonic portion. Such a nozzle is called a convergent-divergent nozzle. A convergent-divergent nozzle is also called a de laval nozzle, after Carl G.P. de Laval who first used such a configuration in his steam turbines in late nineteenth century.

From figure, it is clear that the Mach number must be unity at the throat, where the area is neither increasing nor decreasing. This is consistent with equation $\frac{dA}{A} = -\frac{dV}{V}\left[1 - Ma^2\right]$ which shows that dV can be nonzero at the throat only if Ma =1. It also follows that the sonic velocity can be achieved only at the throat of a nozzle or a diffuser.

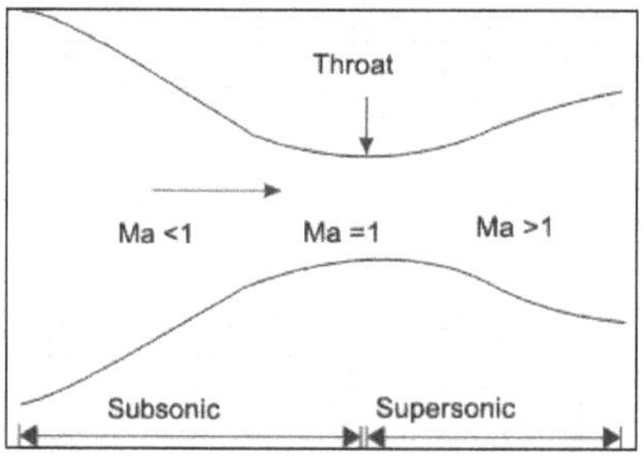

A Convergent-Divergent Nozzle.

The condition, however, does not restrict that Ma must necessarily be unity at the throat. According to equation $\frac{dA}{A} = -\frac{dV}{V}\left[1 - Ma^2\right]$, a situation is possible where Ma ≠ 1 at the throat if dV = othere. For an example, the flow in a convergent-divergent duct may be subsonic everywhere with Ma increasing in the convergent portion and decreasing in the divergent portion with Ma ≠ 1 at the throat.

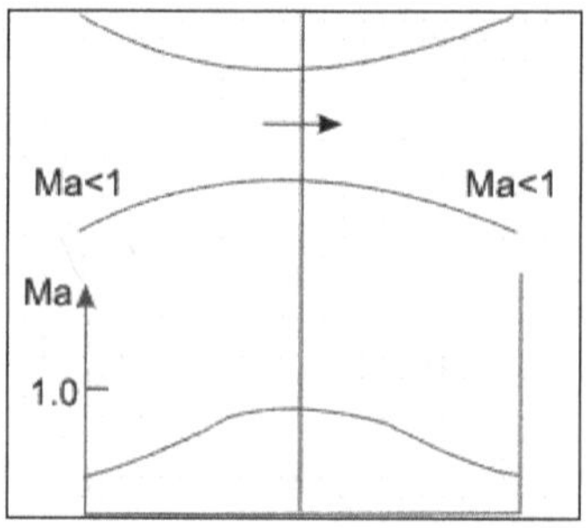

Convergent-Divergent duct with Ma ≠ 1 at throat.

The first part of the duct is acting as a nozzle, whereas the second part is acting as a diffuser. Alternatively, we may have a convergent divergent duct in which the flow is supersonic everywhere with Ma decreasing in the convergent part and increasing in the divergent part and again Ma ≠ 1 at the throat.

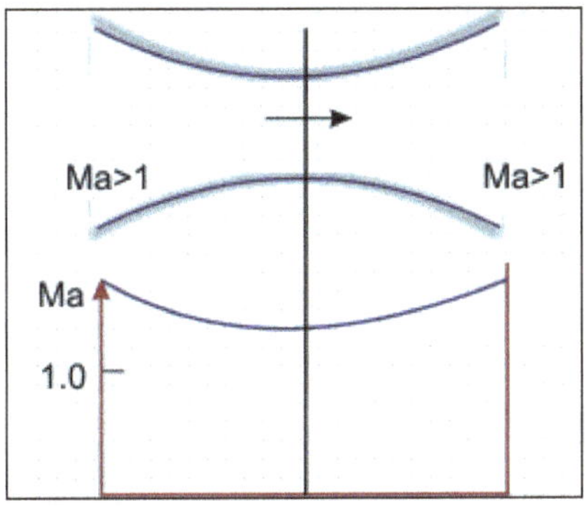

Convergent-Divergent duct with Ma ≠ 1 at throat.

Consider the mass flow rate of an ideal gas through a converging nozzle. If the flow is isentropic, we can write:

$$\dot{m} = \rho A V$$

where V is flow velocity, A is area, ρ is the density of the field.

This can equivalently be written as:

$$\frac{\dot{m}}{A} = \frac{p}{RT} \cdot a \cdot Ma$$

$$\frac{\dot{m}}{A} = \frac{p}{RT} \cdot \sqrt{\gamma RT} \cdot Ma$$

$$\frac{\dot{m}}{A} = \frac{p}{\sqrt{T}} \cdot \sqrt{\frac{\gamma}{R}} \cdot Ma$$

$$\frac{\dot{m}}{A} = \frac{p}{p_0} \cdot \sqrt{\frac{T_0}{T}} \cdot \sqrt{\frac{1}{T_0}} \cdot \sqrt{\frac{\gamma}{R}} \cdot Ma$$

$$\frac{\dot{m}}{A} = \left(\frac{T_0}{T}\right)^{\frac{\gamma}{\gamma-1}} \left(\frac{T_0}{T}\right)^{\frac{1}{2}} \frac{p_0}{\sqrt{T_0}} \cdot \sqrt{\frac{1}{T_0}} \cdot Ma$$

$$\frac{\dot{m}}{A} = \sqrt{\frac{\gamma}{R}} \cdot \frac{p_0 Ma}{\sqrt{T_0}} \cdot \left(\frac{T_0}{T}\right)^{-\frac{\gamma+1}{2(\gamma-1)}}$$

$$\frac{\dot{m}}{A} = \sqrt{\frac{\gamma}{R}} \cdot \frac{p_0 Ma}{\sqrt{T_0}} \cdot \frac{1}{\left[1 + \frac{\gamma-1}{2} Ma^2\right]^{-\frac{\gamma+1}{2(\gamma-1)}}}$$

p_0, T_0, γ and R are constant.

- The discharge per unit area $\frac{\dot{m}}{A}$ is a function of Ma only. There exists a particular value of Ma for which it is maximum. Differentiating with respect to Ma and equating it to zero, we get:

$$\frac{\dot{m}}{A}=\sqrt{\frac{\gamma}{R}}\cdot\frac{p_0}{\sqrt{T_0}}\cdot\frac{1}{\left[1+\frac{\gamma-1}{2}Ma^2\right]^{-\frac{\gamma+1}{2(\gamma-1)}}}$$

$$+\sqrt{\frac{\gamma}{R}}\cdot\frac{p_0 Ma}{\sqrt{T_0}}\left[-\frac{\gamma+1}{2(\gamma-1)}\left\{1+\frac{\gamma-1}{2}Ma^2\right\}\left\{\frac{\gamma+1}{2(\gamma-1)}-1\right\}\left\{\frac{\gamma-1}{2}2Ma\right\}\right]=0$$

$$\Rightarrow 1-\frac{Ma^2(\gamma+1)}{2\left\{1+\frac{\gamma-1}{2}Ma^2\right\}}=0$$

$$\Rightarrow Ma^2(\gamma+1)=2+(\gamma-1)Ma^2$$

$$\Rightarrow Ma=1$$

Hence, discharge is maximum when Ma = 1.

- We know that $V=aMa=\sqrt{rRT}Ma$. By logarithmic differentiation, we get:

$$\frac{dV}{V}=\frac{dMa}{Ma}+\frac{1}{2}\frac{dT}{T}$$

We also know that:

$$\frac{T}{T_0}=\left[1+\frac{\gamma-1}{2}Ma^2\right]^{-1}$$

By logarithmic differentiation, we get:

$$\frac{dT}{T}=-\frac{(\gamma-1)Ma^2}{1+\frac{\gamma-1}{2}Ma^2}\cdot\frac{dMa}{Ma}$$

From equations $\frac{T}{T_0}=\left[1+\frac{\gamma-1}{2}Ma^2\right]^{-1}$ and $\frac{dT}{T}=-\frac{(\gamma-1)Ma^2}{1+\frac{\gamma-1}{2}Ma^2}\cdot\frac{dMa}{Ma}$, we get:

$$\frac{dV}{V}=\frac{dMa}{Ma}\left[1-\frac{\frac{\gamma-1}{2}Ma^2}{1+\frac{\gamma-1}{2}Ma^2}\right]$$

$$\Rightarrow \frac{dV}{V}=\frac{dMa}{Ma}\left[\frac{1}{1+\frac{\gamma-1}{2}Ma^2}\right]$$

From equations $\frac{dA}{A}=-\frac{dV}{V}\left[1-Ma^2\right]$ and $\frac{dV}{V}=\frac{dMa}{Ma}\left[1-\frac{\frac{\gamma-1}{2}Ma^2}{1+\frac{\gamma-1}{2}Ma^2}\right]$ $\Rightarrow \frac{dV}{V}=\frac{dMa}{Ma}\left[\frac{1}{1+\frac{\gamma-1}{2}Ma^2}\right]$ we get:

$$\frac{dA}{A}\frac{1}{\left(Ma^2-1\right)}=\frac{1}{1+\frac{\gamma-1}{2}Ma^2}\frac{dMa}{Ma}$$

$$\frac{dA}{A}=\frac{\left(Ma^2-1\right)}{1+\frac{\gamma-1}{2}Ma^2}\frac{dMa}{Ma}$$

By substituting Ma=1 in equation $\frac{dA}{A}\frac{1}{\left(Ma^2-1\right)}=\frac{1}{1+\frac{\gamma-1}{2}Ma^2}\frac{dMa}{Ma}$ $\frac{dA}{A}=\frac{\left(Ma^2-1\right)}{1+\frac{\gamma-1}{2}Ma^2}\frac{dMa}{Ma}$,we get dA = 0 or A = constant.

Ma=1 can occur only at the throat and nowhere else, and this happens only when the discharge is maximum. When Ma = 1, the discharge is maximum and the nozzle is said to be choked.

$$\frac{\dot{m}}{A^*}=\sqrt{\frac{\gamma}{R}}\cdot\frac{p_0}{\sqrt{T_0}}\cdot\frac{1}{\left[\left(\frac{\gamma+1}{2}\right)\right]^{\frac{\gamma+1}{2(\gamma-1)}}}$$

(as we have earlier designated critical or sonic conditions by a superscript asterisk). Dividing equation:

$$\frac{\dot{m}}{A}=\frac{p}{RT}\cdot a\cdot Ma$$

$$\frac{\dot{m}}{A}=\frac{p}{RT}\cdot\sqrt{\gamma RT}\cdot Ma$$

$$\frac{\dot{m}}{A}=\frac{p}{\sqrt{T}}\cdot\sqrt{\frac{\gamma}{R}}\cdot Ma$$

$$\frac{\dot{m}}{A}=\frac{p}{p_0}\cdot\sqrt{\frac{T_0}{T}}\cdot\sqrt{\frac{1}{T_0}}\cdot\sqrt{\frac{\gamma}{R}}\cdot Ma$$

$$\frac{\dot{m}}{A}=\left(\frac{T_0}{T}\right)^{\frac{\gamma}{\gamma-1}}\left(\frac{T_0}{T}\right)^{\frac{1}{2}}\frac{p_0}{\sqrt{T_0}}\cdot\sqrt{\frac{1}{T_0}}\cdot Ma$$

$$\frac{\dot{m}}{A}=\sqrt{\frac{\gamma}{R}}\cdot\frac{p_0 Ma}{\sqrt{T_0}}\cdot\left(\frac{T_0}{T}\right)^{-\frac{\gamma+1}{2(\gamma-1)}}$$

$\frac{\dot{m}}{A^*}=\sqrt{\frac{\gamma}{R}}\cdot\frac{p_0}{\sqrt{T_0}}\cdot\frac{1}{\left[\left(\frac{\gamma+1}{2}\right)\right]^{-\frac{\gamma+1}{2(\gamma-1)}}}$ by Eq. $\frac{\dot{m}}{A}=\sqrt{\frac{\gamma}{R}}\cdot\frac{p_0 Ma}{\sqrt{T_0}}\cdot\frac{1}{\left[1+\frac{\gamma-1}{2}Ma^2\right]^{-\frac{\gamma+1}{2(\gamma-1)}}}$ we obtain:

$$\frac{A}{A^*}=\frac{1}{Ma}\left[\left\{\frac{2}{\gamma+1}\right\}\left\{1+\frac{(\gamma-1)}{2}Ma^2\right\}\right]\frac{\gamma+1}{2(\gamma-1)}$$

From equation $\frac{A}{A^*}=\frac{1}{Ma}\left[\left\{\frac{2}{\gamma+1}\right\}\left\{1+\frac{(\gamma-1)}{2}Ma^2\right\}\right]\frac{\gamma+1}{2(\gamma-1)}$ we see that a choice of Ma gives a unique value of A/A*. The following figure shows variation of A / A * with Ma. Note that the curve is double valued; that is, for a given value of A/A* (other than unity), there are two possible values of Mach number. This signifies the fact that the supersonic nozzle is diverging.

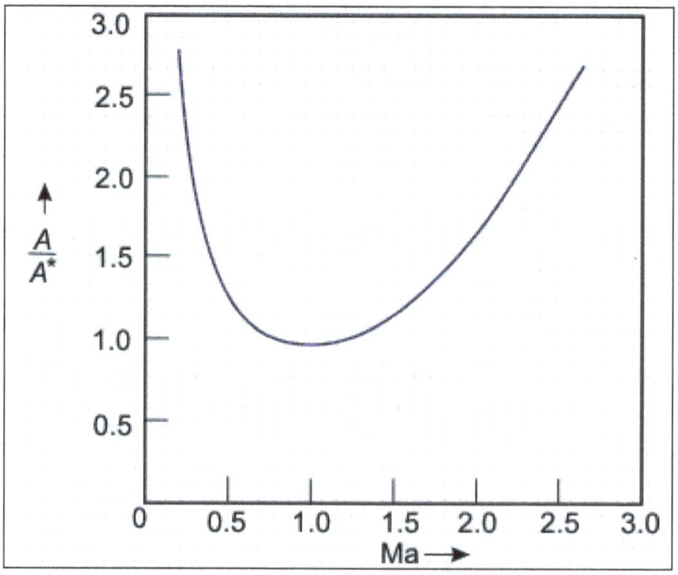

Variation of A/A* with Ma in isentropic flow for γ = 1.4.

Pressure Distribution and Choking in a Converging Nozzle

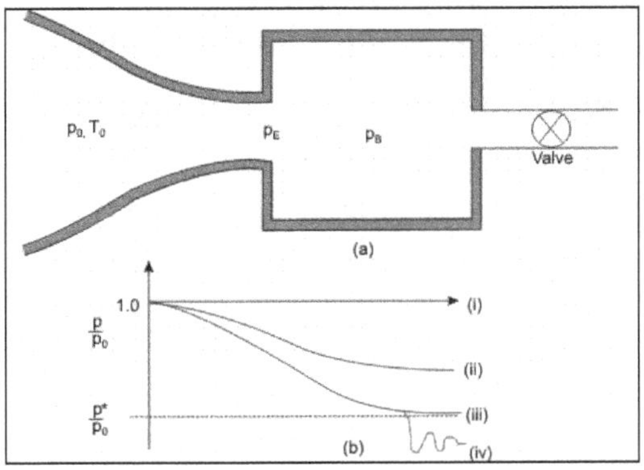

(a) Compressible flow through a converging nozzle (b) Pressure distribution along a converging nozzle for different values of back pressure.

- Consider a convergent nozzle as shown in figure (a) and (b) shows the pressure ratio p/p_o along the length of the nozzle.
- The inlet conditions of the gas are at the stagnation state (p_o, T_o) which are constants. The pressure at the exit plane of the nozzle is denoted by P_E and the back pressure is P_B which can be varied by the adjustment of the valve. At the condition $P_o = P_E = P_B$ there shall be no flow through the nozzle.
- The pressure is P_0 throughout, as shown by condition (i) in figure (b). As P_B is gradually reduced, the flow rate shall increase. The pressure will decrease in the direction of flow as shown by condition (ii) in figure (b). The exit plane pressure P_E shall remain equal to P_B so long as the maximum discharge condition is not reached. Condition (iii) in figure (b) illustrates the pressure distribution in the maximum discharge situation.
- When $\frac{\dot{m}}{A}$ attains its maximum value, given by substituting Ma = 1 in equation:

$$\frac{\dot{m}}{A} = \frac{p}{RT} \cdot a \cdot Ma$$

$$\frac{\dot{m}}{A} = \frac{p}{RT} \cdot \sqrt{\gamma RT} \cdot Ma$$

$$\frac{\dot{m}}{A} = \frac{p}{\sqrt{T}} \cdot \sqrt{\frac{\gamma}{R}} \cdot Ma$$

$$\frac{\dot{m}}{A} = \frac{p}{p_0} \cdot \sqrt{\frac{T_0}{T}} \cdot \sqrt{\frac{1}{T_0}} \cdot \sqrt{\frac{\gamma}{R}} \cdot Ma$$

$$\frac{\dot{m}}{A} = \left(\frac{T_0}{T}\right)^{\frac{\gamma}{\gamma-1}} \left(\frac{T_0}{T}\right)^{\frac{1}{2}} \frac{p_0}{\sqrt{T_0}} \cdot \sqrt{\frac{1}{T_0}} \cdot Ma$$

$$\frac{\dot{m}}{A} = \sqrt{\frac{\gamma}{R}} \cdot \frac{p_0 Ma}{\sqrt{T_0}} \cdot \left(\frac{T_0}{T}\right)^{-\frac{\gamma+1}{2(\gamma-1)}}$$

$$\frac{\dot{m}}{A} = \sqrt{\frac{\gamma}{R}} \cdot \frac{p_0 Ma}{\sqrt{T_0}} \cdot \frac{1}{\left[1 + \frac{\gamma-1}{2} Ma^2\right]^{\frac{\gamma+1}{2(\gamma-1)}}}$$

P_E is equal to p*. Since the nozzle does not have a diverging section, further reduction in back pressure P_B will not accelerate the flow to supersonic condition. As a result, the exit pressure P_E shall continue to remain at p * even though P_B is lowered further.

- The convergent-nozzle discharge against the variation of back pressure is shown in figure. We are aware, that the maximum value of (m/A) at Ma = 1 is stated as the choked flow. With a given nozzle, the flow rate cannot be increased further. Thus neither the nozzle exit pressure, nor the mass flow rate are affected by lowering P_B below p *.

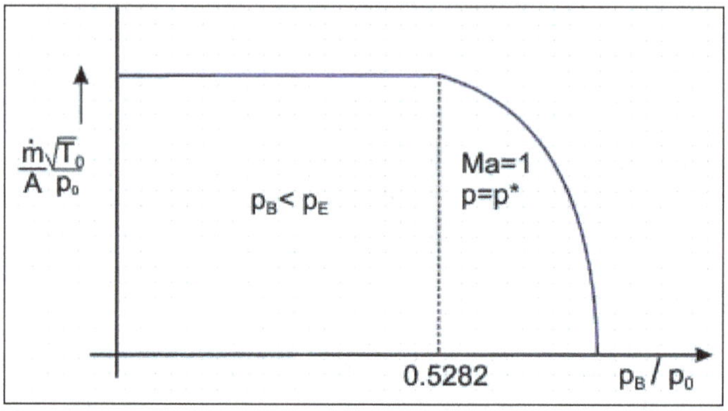

Mass flow rate and the variation of back pressure in a converging nozzle.

However for P_B less than p *, the flow leaving the nozzle has to expand to match the lower back pressure as shown by condition (iv) in figure (b). This expansion process is three-dimensional and the pressure distribution cannot be predicted by one-dimensional theory. Experiments reveal that a series of shocks form in the exit stream, resulting in an increase in entropy.

Isentropic Flow in a Converging-Diverging Nozzle

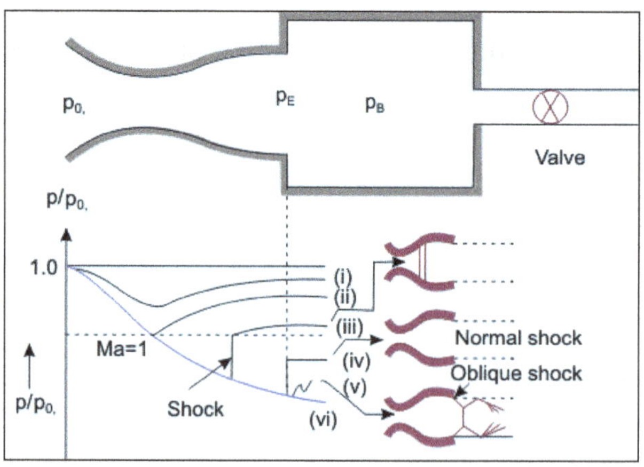

Pressure Distribution along a Converging-Diverging Nozzle for different values of back pressure P_B.

- Consider the flow in a convergent-divergent nozzle. The upstream stagnation conditions are assumed constant; the pressure in the exit plane of the nozzle is denoted by P_E ; the nozzle discharges to the back pressure, P_B .

- With the valve initially closed, there is no flow through the nozzle; the pressure is constant at P_0. Opening the valve slightly produces the pressure distribution shown by curve (i). Completely subsonic flow is discerned.

- Then P_B is lowered in such a way that sonic condition is reached at the throat (ii). The flow rate becomes maximum for a given nozzle and the stagnation conditions.
- On further reduction of the back pressure, the flow upstream of the throat does not respond. However, if the back pressure is reduced further (cases (iii) and (iv)), the flow initially becomes supersonic in the diverging section, but then adjusts to the back pressure by means of a normal shock standing inside the nozzle. In such cases, the position of the shock moves downstream as P_B is decreased, and for curve (iv) the normal shock stands right at the exit plane.
- The flow in the entire divergent portion up to the exit plane is now supersonic. When the back pressure is reduced even further (v), there is no normal shock anywhere within the nozzle, and the jet pressure adjusts to P_B by means of oblique shock waves outside the exit plane. A converging diverging nozzle is generally intended to produce supersonic flow near the exit plane.
- If the back pressure is set at (vi), the flow will be isentropic throughout the nozzle, and supersonic at nozzle exit. Nozzles operating at P_B (corresponding to curve (vi) in figure) are said to be at design conditions. Rocket-propelled vehicles use converging-diverging nozzles to accelerate the exhaust gases to the maximum possible velocity to produce high thrust.

Steam Injector

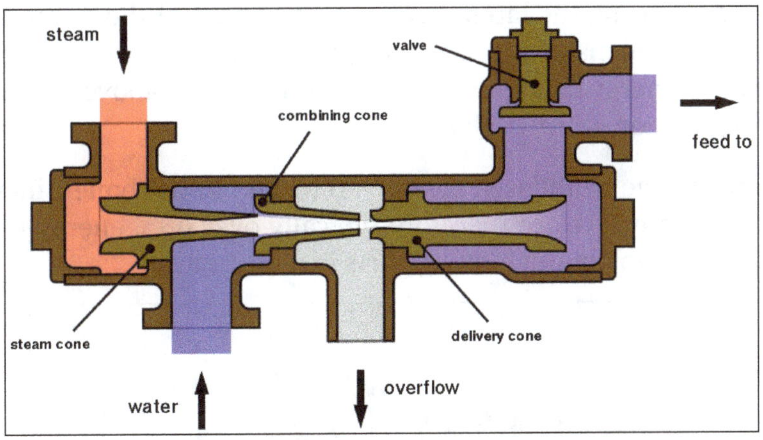

Injector used in steam locomotives.

A steam injector is typically used to deliver cold water to a boiler against its own pressure using its own live or exhaust steam, replacing any mechanical pump. This was the purpose for which it was originally invented in 1858 by Henri Giffard. Its operation was from the start intriguing since it seemed paradoxical, almost like perpetual motion, but its operation was later explained using thermodynamics. Other types of injector may use other pressurised motive fluids such as air.

Depending on the application, an injector can also take the form of an eductor-jet pump, a water eductor or an aspirator. An ejector operates on similar principles to create a vacuum feed connection for braking systems etc.

Operation

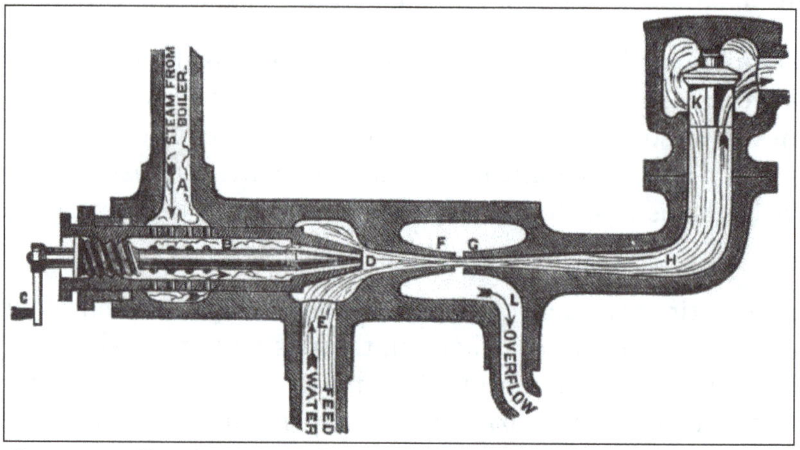

A- Steam from boiler, B- Needle valve, C- Needle valve handle, D- Steam and water combine, E- Water feed, F- Combining cone, G- Delivery nozzle and cone, H- delivery chamber and pipe, K- Check valve, L- Overflow.

The injector consists of a body filled with a secondary fluid, into which a motive fluid is injected. The motive fluid induces the secondary fluid to move. Injectors exist in many variations, and can have several stages, each repeating the same basic operating principle, to increase their overall effect.

It uses the Venturi effect of a converging-diverging nozzle on a steam jet to convert the pressure energy of the steam to velocity energy, reducing its pressure to below atmospheric which enables it to entrain a fluid (eg. water). After passing through the convergent "combining cone", the mixed fluid is fully condensed releasing the latent heat of evaporation of the steam which imparts extra velocity to the water. The condensate mixture then enters a divergent "delivery cone" which slows the jet, converting kinetic energy back into static pressure energy above the pressure of the boiler enabling its feed through a non-return valve.

Most of the heat energy in the condensed steam is returned to the boiler, increasing the thermal efficiency of the process. Injectors are therefore typically over 98% energy-efficient overall; they are also simple compared to the many moving parts in a feed pump.

Steam injector of a locomotive boiler.

The motive fluid may be a liquid, steam or any other gas. The entrained suction fluid may be a gas, a liquid, a slurry, or a dust-laden gas stream.

Design Parameters

Fluid feed rate and operating pressure range are the key parameters of an injector, and vacuum pressure and evacuation rate for an ejector.

Compression ratio and the entrainment ratio may also be defined:

The compression ratio of the injector, P_2 / P_1, is defined as ratio of the injector's outlet pressure P_2 to the inlet pressure of the suction fluid P_1.

The entrainment ratio of the injector, W_s / W_m, is defined as the amount W_s (in kg/h) of suction fluid that can be entrained and compressed by a given amount W_m (in kg/h) of motive fluid.

Lifting Properties

Other key properties of an injector include the fluid inlet pressure requirements i.e. whether it is lifting or non-lifting.

In a non-lifting injector, positive inlet fluid pressure is needed e.g. the cold water-input is fed by gravity.

The steam-cone minimal orifice diameter is kept larger than the combining cone minimal diameter. The non-lifting Nathan 4000 injector used on the Southern Pacific 4294 could push 12,000 US gallons (45,000 L) per hour at 250 psi (17 bar).

The lifting injector can operate with negative inlet fluid pressure i.e. fluid lying below the level of the injector. It differs from the non-lifting type mainly in the relative dimensions of the nozzles.

Overflow

An overflow is required for excess steam or water to discharge, especially during starting; if the injector cannot initially overcome boiler pressure, the overflow allows the injector to continue to draw water and steam.

Check Valve

There is at least one check valve (called a "clack valve" in locomotives because of the distinctive noise it makes) between the exit of the injector and the boiler to prevent back flow, and usually a valve to prevent air being sucked in at the overflow.

Exhaust Steam Injector

Efficiency was further improved by the development of a multi-stage injector which is powered not by live steam from the boiler but by exhaust steam from the cylinders, thereby making use of the residual energy in the exhaust steam which would otherwise have gone to waste. However, an exhaust injector also cannot work when the locomotive is stationary; later exhaust injectors could use a supply of live steam if no exhaust steam was available.

Problems

Injectors can be troublesome under certain running conditions, when vibration caused the combined steam and water jet to "knock off". Originally the injector had to be restarted by careful manipulation of the steam and water controls, and the distraction caused by a malfunctioning injector was largely responsible for the 1913 Ais Gill rail accident. Later injectors were designed to automatically restart on sensing the collapse in vacuum from the steam jet, for example with a spring-loaded delivery cone.

Another common problem occurs when the incoming water is too warm and is less effective at condensing the steam in the combining cone. This can also occur if the metal body of the injector is too hot, e.g. from prolonged use.

Vacuum Ejectors

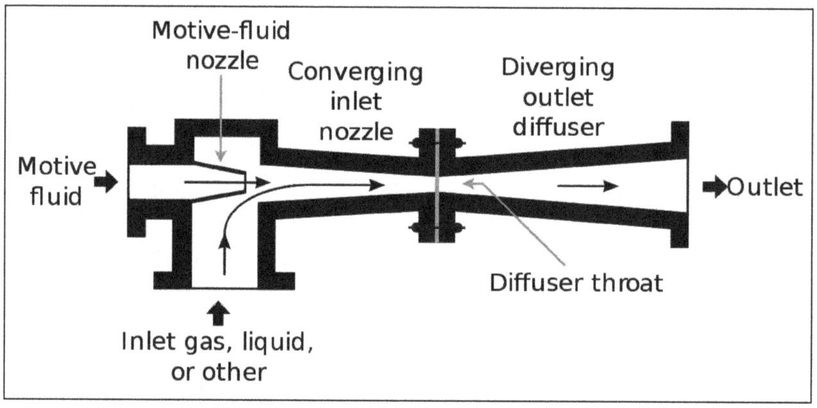

Diagram of a typical modern ejector.

An additional use for the injector technology is in vacuum ejectors in continuous train braking systems, which were made compulsory in the UK by the Regulation of Railways Act 1889. A vacuum ejector uses steam pressure to draw air out of the vacuum pipe and reservoirs of continuous train brake. Steam locomotives, with a ready source of steam, found ejector technology ideal with its rugged simplicity and lack of moving parts. A steam locomotive usually has two ejectors: a large ejector for releasing the brakes when stationary and a small ejector for maintaining the vacuum against leaks. The small ejector is sometimes replaced by a reciprocating pump driven from the crosshead because this is more economical of steam.

Vacuum brakes have been superseded by air brakes in modern trains, which use pumps, as diesel and electric locomotives no longer have a suitable working fluid for vacuum ejectors.

Earlier Application of the Principle

An empirical application of the principle was in widespread use on steam locomotives before its formal development as the injector, in the form of the arrangement of the blastpipe and chimney in the locomotive smokebox. The sketch on the right shows a cross section through a smokebox, rotated 90 degrees; it can be seen that the same components are present, albeit differently named, as in the generic diagram of an injector. Exhaust steam from the cylinders is directed through a nozzle on the end of the blastpipe, to reduce pressure inside the smokebox by entraining the flue gases from the boiler which are then ejected via the chimney. The effect is to increase the draught on the fire to a degree proportional to the rate of steam consumption, so that as more steam is used, more heat is generated from the fire and steam production is also increased. The effect was first noted by Richard Trevithick and subsequently developed empirically by the early locomotive engineers; Stephenson's Rocket made use of it, and this constitutes much of the reason for its notably improved performance in comparison with contemporary machines.

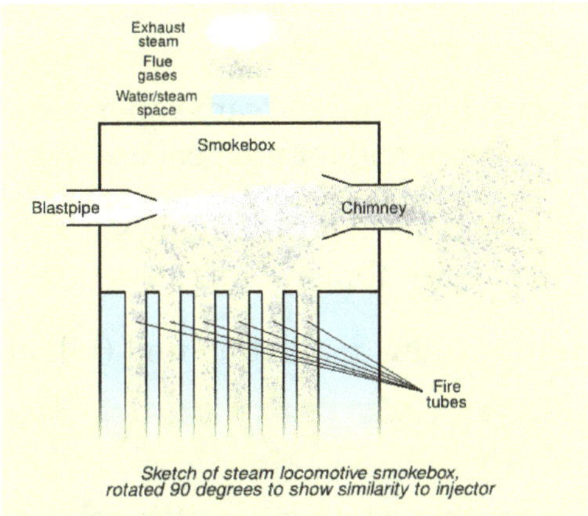

Sketch of the smokebox of a steam locomotive, rotated 90 degrees. The similarity to the generic injector diagram at the top of this article is apparent.

Modern Uses

The use of injectors (or ejectors) in various industrial applications has become quite common due to their relative simplicity and adaptability. For example:

- To inject chemicals into the boiler drums of small, stationary, low pressure boilers. In large, high-pressure modern boilers, usage of injectors for chemical dosing is not possible due to their limited outlet pressures.
- In thermal power stations, they are used for the removal of the boiler bottom ash, the removal of fly ash from the hoppers of the electrostatic precipitators used to remove that ash from the boiler flue gas, and for drawing a vacuum pressure in steam turbine exhaust condensers.
- Jet pumps have been used in boiling water nuclear reactors to circulate the coolant fluid.
- For use in producing a vacuum pressure in steam jet cooling systems.
- For expansion work recovery in air conditioning and refrigeration systems.
- For enhanced oil recovery processes in the oil & gas Industry.
- For the bulk handling of grains or other granular or powdered materials.
- The construction industry uses them for pumping turbid water and slurries.
- Eductors are used in ships to pump residual ballast water, or cargo oil which cannot be removed using centrifugal pumps due to loss of suction head and may damage the centrifugal pump if run dry, which may be caused due to trim or list of the ship.
- Eductors are used on-board ships to pump out bilges, since using centrifugal pump would not be feasible as the suction head may be lost frequently.

- Some aircraft (mostly earlier designs) use an ejector attached to the fuselage to provide vacuum for gyroscopic instruments such as an attitude indicator (artificial horizon).
- Eductors are used in aircraft fuel systems as transfer pumps; fluid flow from an engine-mounted mechanical pump can be delivered to a fuel tank-mounted eductor to transfer fuel from that tank.
- Aspirators are vacuum pumps based on the same operating principle and are used in laboratories to create a partial vacuum and for medical use in suction of mucus or bodily fluids.
- Water eductors are water pumps used for dredging silt and panning for gold, they're used because they can handle the highly abrasive mixtures quite well.
- To create vacuum system in vacuum distillation unit (oil refinery).

Well Pumps

Jet pumps are commonly used to extract water from water wells. The main pump, often a centrifugal pump, is powered and installed at ground level. Its discharge is split, with the greater part of the flow leaving the system, while a portion of the flow is returned to the jet pump installed below ground in the well. This recirculated part of the pumped fluid is used to power the jet. At the jet pump, the high-energy, low-mass returned flow drives more fluid from the well, becoming a low-energy, high-mass flow which is then piped to the inlet of the main pump.

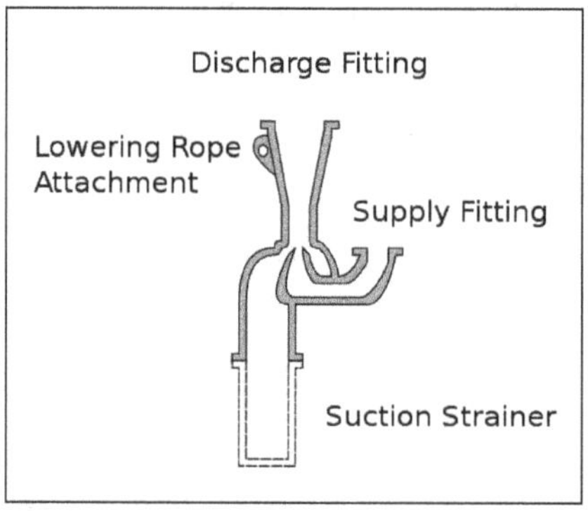

The S type pump is useful for removing water from a well or container.

Shallow well pumps are those in which the jet assembly is attached directly to the main pump and are limited to a depth of approximately 5-8m to prevent cavitation.

Deep well pumps are those in which the jet is located at the bottom of the well. The maximum depth for deep well pumps is determined by the inside diameter of and the velocity through the jet. The major advantage of jet pumps for deep well installations is the ability to situate all mechanical parts (e.g., electric/petrol motor, rotating impellers) at the ground surface for easy maintenance. The advent of the electrical submersible pump has partly replaced the need for jet type well pumps, except for driven point wells or surface water intakes.

Multi-stage Steam Vacuum Ejectors

In practice, for suction pressure below 100 mbar absolute, more than one ejector is used, usually with condensers between the ejector stages. Condensing of motive steam greatly improves ejector set efficiency; both barometric and shell-and-tube surface condensers are used.

In operation a two-stage system consists of a primary high-vacuum (HV) ejector and a secondary low-vacuum (LV) ejector. Initially the LV ejector is operated to pull vacuum down from the starting pressure to an intermediate pressure. Once this pressure is reached, the HV ejector is then operated in conjunction with the LV ejector to finally pull vacuum to the required pressure.

In operation a three-stage system consists of a primary booster, a secondary high-vacuum (HV) ejector, and a tertiary low-vacuum (LV) ejector. As per the two-stage system, initially the LV ejector is operated to pull vacuum down from the starting pressure to an intermediate pressure. Once this pressure is reached, the HV ejector is then operated in conjunction with the LV ejector to pull vacuum to the lower intermediate pressure. Finally the booster is operated (in conjunction with the HV & LV ejectors) to pull vacuum to the required pressure.

Permissions

All chapters in this book are published with permission under the Creative Commons Attribution Share Alike License or equivalent. Every chapter published in this book has been scrutinized by our experts. Their significance has been extensively debated. The topics covered herein carry significant information for a comprehensive understanding. They may even be implemented as practical applications or may be referred to as a beginning point for further studies.

We would like to thank the editorial team for lending their expertise to make the book truly unique. They have played a crucial role in the development of this book. Without their invaluable contributions this book wouldn't have been possible. They have made vital efforts to compile up to date information on the varied aspects of this subject to make this book a valuable addition to the collection of many professionals and students.

This book was conceptualized with the vision of imparting up-to-date and integrated information in this field. To ensure the same, a matchless editorial board was set up. Every individual on the board went through rigorous rounds of assessment to prove their worth. After which they invested a large part of their time researching and compiling the most relevant data for our readers.

The editorial board has been involved in producing this book since its inception. They have spent rigorous hours researching and exploring the diverse topics which have resulted in the successful publishing of this book. They have passed on their knowledge of decades through this book. To expedite this challenging task, the publisher supported the team at every step. A small team of assistant editors was also appointed to further simplify the editing procedure and attain best results for the readers.

Apart from the editorial board, the designing team has also invested a significant amount of their time in understanding the subject and creating the most relevant covers. They scrutinized every image to scout for the most suitable representation of the subject and create an appropriate cover for the book.

The publishing team has been an ardent support to the editorial, designing and production team. Their endless efforts to recruit the best for this project, has resulted in the accomplishment of this book. They are a veteran in the field of academics and their pool of knowledge is as vast as their experience in printing. Their expertise and guidance has proved useful at every step. Their uncompromising quality standards have made this book an exceptional effort. Their encouragement from time to time has been an inspiration for everyone.

The publisher and the editorial board hope that this book will prove to be a valuable piece of knowledge for students, practitioners and scholars across the globe.

Index

www.ingramcontent.com/pod-product-compliance
Lightning Source LLC
LaVergne TN
LVHW080721210926
848485LV00003B/499